STUDY GUIDE

Lydia Daniels & Laurel Bridges Roberts
University of Pittsburgh

BIOLOGY

A GUIDE TO THE NATURAL WORLD

David Krogh

Third Edition

Upper Saddle River, NJ 07458

Project Manager: Crissy Dudonis
Executive Editor: Gary Carlson
Editor-in-Chief, Science: John Challice
Vice President of Production & Manufacturing: David W. Riccardi
Executive Managing Editor: Kathleen Schiaparelli
Assistant Managing Editor: Becca Richter
Production Editor: Dana Dunn
Supplement Cover Manager: Paul Gourhan
Supplement Cover Designer: Joanne Alexandris
Manufacturing Buyer: Ilene Kahn

Pearson Prentice Hall
Pearson Education, Inc.
Upper Saddle River, NJ 07458

Printed in the United States of America

10 9 8 7 6 5 4

ISBN 0-13-144934-6

Pearson Education Ltd., *London*
Pearson Education Australia Pty. Ltd., *Sydney*
Pearson Education Singapore, Pte. Ltd.
Pearson Education North Asia Ltd., *Hong Kong*
Pearson Education Canada, Inc., *Toronto*
Pearson Educación de Mexico, S.A. de C.V.
Pearson Education—Japan, *Tokyo*
Pearson Education Malaysia, Pte. Ltd.

Contents

PREFACE
Welcome to the Study Guide

Welcome to the *Study Guide*, a companion for David Krogh's *Biology: A Guide to the Natural World.* Biology is the study of the world around us: how organisms live, eat, move, reproduce, and change over time. Biological knowledge is increasing at a dizzying pace, and that pace shows no indication of slowing down. Biological issues —including cloning, drug design, environmental management, and evolutionary processes—influence both political and sociological issues. This situation makes biologic literacy a critical component of good citizenship in our nation and our world. Understanding biology will help you understand the world around you and your place within it.

We are sure you want to be as successful as possible in your study of biology. Luckily, there are plenty of resources for you. In addition to your instructor, your lecture notes, your textbook, and your classmates, we have assembled this study guide for you. Remember that this guide is only one possible resource. It is not meant to replace your lecture notes or your textbook. Use this guide to reinforce and test your understanding of the concepts and details presented in class.

Our study guide is designed to help you identify and learn the important concepts presented in your textbook. Every *Study Guide* chapter has a similar format:

- A summary of the chapter's three or four **Basic Concepts**
- A bulleted list of the **key ideas** related to each of the Basic Concepts
  This list should help you identify the big ideas that you need to understand thoroughly to master the information in each chapter
- **Think About It** – a list of questions to test how well you've grasped the basic concepts.
- A **Cool Facts** essay presenting current research or real-life applications of the material presented in the chapter
- A **Vocabulary Review** for the entire chapter
- Suggested terms for **Web building** and **Flow charting**
- A **Practice Test**
- A practice essay question, called **But, What's It all About**, that challenges you to use the information you have learned not only in the current chapter but also in previous chapters to solve a problem or defend a position.

You can read and take notes on your text and then use the Basic Concepts and bulleted lists to double check your grasp of the major ideas in each chapter. To test your comprehension of the factual content, you can work through the Think About It exercises in the *Study Guide* as you finish reading the relevant section of the textbook chapter. Building the suggested webs and flowcharts will help you see how the factual information presented in the textbook defines the Basic Concepts of the chapter. When you are familiar with the material, you can count on the vocabulary review and practice test at the end of the chapter to assess your strengths and identify your weaknesses. You can also use the vocabulary review, practice test, and But, What's It All About question to prepare for your exams. The But, What's It All About question is structured to teach you how to collect evidence from multiple sources and use it to write a clear and detailed answer to a question – an important skill in any discipline. Even if your instructor doesn't use essay questions on tests, this practice essay will teach you how to integrate your knowledge of the factual content of biology and develop a deep and long-lasting understanding of the material.

We hope you will learn to love biology as much as we do! The understanding you gain from your studies will bring you a lifelong appreciation for the diversity and beauty of our world. We would like to acknowledge the work of our friend and colleague, Dr. Michele Shuster, whose creative energy and good humor contributed immeasurably to this *Study Guide*. Finally, we would like to thank our mentors for guiding us to become better scientists, our students for guiding us to become better teachers, and our families for guiding us to become kinder (and calmer) human beings.

Lydia Daniels

Laurel Roberts

CHAPTER 1

Science as a Way of Learning: A Guide to the Natural World

Hello! We (the authors of this *Study Guide*) know who you are.

Well, we have a pretty good idea about who you are. You are enrolled in a biology class somewhere that is using David Krogh's *Biology: A Guide to the Natural World* as a textbook. Maybe you are enrolled in the class because you have to fill a distribution requirement for a natural science, and biology seemed less threatening than chemistry or physics. Maybe you want to learn some basic biology because it would help you understand the latest developments in the field of law, business, education, or communications. Maybe it's been a long time since you studied biology and you're curious about the science behind vitamins, global warming, and evolution. Whatever your reasons for learning about biology, we thought you'd like to know what your instructor will expect you to learn.

We have a pretty good idea about who your instructor is, too. Probably someone like us, who loves studying the natural world and enjoys sharing their knowledge of biology with other people. They may also do biological research and have specialized knowledge about some aspect of biology (Laurel is an ecologist; Lydia is a biochemist). But your instructor knows that biology isn't your primary field of interest. This is one of many courses you may be enrolled in, and it may not even be your favorite. So what does he or she expect you to learn?

We can't answer that question specifically, because every instructor's expectations are unique. But we can show you some learning tools that will help you deal with a variety of teaching and testing styles. In the first chapter of the text, David Krogh eloquently describes why you should learn biology; we want to use this chapter to show you how to learn biology—or any other complex subject, for that matter. We'll be introducing the techniques of concept mapping (also called webbing), summary, flowcharting, and grouping. These techniques will help you figure out the basic *concepts* in biology (webbing, summary), the *process* by which we understand these concepts (summary, flowcharting), and how to *apply* specific information to demonstrate these processes (flowcharting, grouping). Using this sequence of concept-process-application will help you learn and remember specific facts, as well as how these facts relate to each other and define biological theories. This is what we expect our students to learn, and we suspect your instructor does too.

Whatever your reasons for learning biology, we hope you'll also spend some time with us in learning how to learn. These techniques not only help you get the greatest value from this biology course but also transfer readily to any discipline, and they'll help you be more successful in learning history, economics, or physics. The knowledge you gain will be uniquely yours—and that's something you can't lose, even if you lose your notes and sell your textbook. So, let's begin.

WHAT IS SCIENCE?

This part of the chapter describes some basic information about how to gather information in science and defines a group of terms commonly used by scientists. Let's begin with a summary outline to record the main idea of this section and summarize the major related concepts.

Science is both a body of knowledge and a method to gain that knowledge.

- Theories are statements of general principles about the natural world supported by evidence.
- Scientific evidence consists of observations and the results of experiments done to account for the observations.
- Experiments proceed using the scientific method, a process that involves testing the validity of a proposed explanation (hypothesis) in a controlled manner.
- No hypothesis or theory is ever proven absolutely in science, but if we fail to disprove theories often enough, we believe them to be true.

Notice that this summary doesn't just copy phrases from the textbook. Each idea has been expressed in terms that make sense to us; that's an important part of making knowledge your own. In this *Study Guide*, you'll see that we often use this summary outline technique to help you find key ideas. Notice also that this summary is rather "bare-bones"—there are no detailed examples of these concepts and no connection between these ideas; it's just a list. To make sense of the details and connections, we need a few other study techniques.

Let's now use a web to establish how these ideas are related and to practice using this new vocabulary.

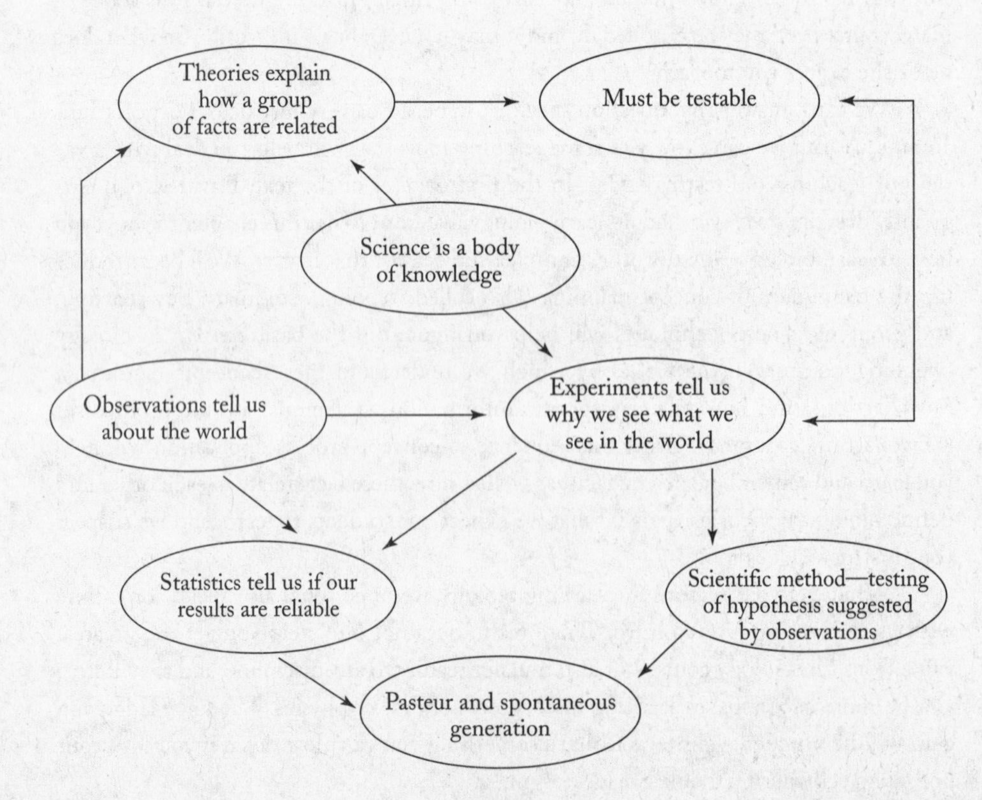

One benefit of using a web is to see how ideas are related; for example, you can see that "theories" are part of the body of knowledge that we call science, and that experiments make up part of that body of knowledge but are also a key part of the process of science (testing theories). Now it's your turn. Use the following list of terms (or make up one of your own) to practice building a concept web.

Create a web: zoo, peanuts, popcorn, elephants, kids, jelly, cartoons, Dumbo, Africa

Webs are particularly useful for showing connections between concepts and processes, but they don't give us any information about the sequence of these connections. For that information, we need to use flowcharts.

Figures 1.5 and 1.6 in the text are examples of *flowcharts*. A flowchart shows the physical or temporal relationships among ideas of things, so it is a good method when you need to remember the order of specific facts. Let's use a flowchart diagram like the one in Figure 1.5 to trace the specifics of Pasteur's experiment in spontaneous generation.

Using the flowchart lets us map the specifics of the Pasteur experiment, an *application* of the scientific method, to the steps of the *process*.

We've demonstrated summary outline, webbing, and flowcharting. Now you can apply these techniques, with our guidance, to the next section, Biology. We'll also introduce you to the last study technique, grouping, using this section. Let's start with a summary outline:

Microbes, living organisms, seem to appear spontaneously in sterile solutions.
(Observation)

↓

Do these microbes arise from chemicals in the solutions, or must there already be microbes present?
(Question)

↓

Microbes are present in the air, so a sterile solution open to the air will grow microbes; but a solution that microbes cannot enter will not.
(Hypothesis)

↓

Broth in swan-necked flasks remains sterile until the neck is broken before the bend, and then microbial life appears in the flask.
(Experiment)

↓

Microbial life appears only when something in the air, but not air alone, can reach the broth. That something must be microbes carried on dust. (Conclusion)

Biology

- Biology is the study of highly organized, living things.
- Living things can be differentiated from nonliving things by a suite of characteristics—there is no single factor that differentiates living from nonliving.
- Living things are highly organized, and this organization occurs in a hierarchy such that relatively simple and small components become integrated to make larger, more complex components.

One general concept mentioned in the outline concerns distinguishing living and nonliving things—what is the process we use to do this? If you read the subsection entitled "What Is Life?" on page 12 of the text, you'll learn some of the processes that characterize living things, but can some of these processes also characterize nonliving things? Here's where you can use *grouping* to help you compare and contrast living and nonliving things.

Grouping allows you to see which characteristics are shared by living and nonliving things and which are unique to living things.

Living Things	Nonliving Things
Movement	Movement (wind, sand dunes)
Growth	Growth (fire, crystals)
Highly organized	Highly organized (crystals, building)
Assimilate and use energy	Use energy (cars)
Respond to the environment	
Maintain internal homeostasis	
Uses an internal information base to function	Uses an internal information base to function (computers)
Can reproduce using internal information base	
Composed of cells	
Evolves from other living things	

The summary outline also indicates that living things are "highly organized." Organized how? In a hierarchy, but a hierarchy of what—and in what order? Sounds like a flowchart is in order. Read the section entitled "Life Is Highly Organized, in a Hierarchical Manner" on page 13, and create a flowchart that shows the components of living things at each level of complexity.

Summary outlines, webs, flowcharts, and groupings are ways to extract information from a text. These techniques give you the ability to explore connections between concepts, the processes by which we know a concept, and the specific details that show the application of a process to a concept. But these techniques are only learning tools, learning occurs when you can use the information the tools have helped you find to explain things. So we present our final study technique, answering questions. You'll see this technique throughout the study guide in sections called "Think about It."

THINK ABOUT IT

1. What kinds of policy issues have you heard about that require a basic knowledge of biology?

2. What is a scientific theory?

3. What are three characteristics of living organisms?

   Are these three characteristics sufficient to define something as living?

4. What is the unifying thread of biology?

Compare and Contrast

For each of the following paired terms, use a sentence of comparison ("Both . . . ") and a sentence of contrast ("However, . . . ").
organelle/organ

cell/organisms

population/community

atoms/molecules

If your instructor has assigned Chapter 1 in the text, you'll want to try the following practice test. Each of the upcoming Study Guide chapters uses a similar formula to help you to prepare for exams.

TESTING, TESTING

After reading Chapter 1 of the text and completing the exercises in Chapter 1 of this Study Guide, try the review exercises and practice test that follow.

Vocabulary Review

Group the following terms in the appropriate columns of the table. Some terms may be used in more than one column.

atom evolution molecular biology scientific method

cell homeostasis organ theory

community hypothesis physiology variable

Hierarchy of Life	Practice of Science	Biology

PRACTICE TEST

After you have reviewed this chapter, close your books, grab a pencil, and spend the next 15 to 20 minutes completing this practice test.

Matching

Match the following terms with their description. Each choice may be used once, more than once, or not at all.

a. the basic unit of matter
b. unified set of principles explaining something about the natural world
c. building block of life
d. gradual change and separation of populations into new species over time

___Evolution
___Theory
___Cell
___Atom

Short Answer

1. What is the relationship between a hypothesis and a theory?

2. Why must a hypothesis be falsifiable?

3. Why is it necessary to define life by a list of criteria instead of a single attribute?

4. How did the study of biology change between the years 1800 and 1900?

Multiple Choice

1. What is the possible explanation for a series of observations?

   a. guess
   b. theory
   c. law
   d. hypothesis
   e. hunch

2. _____ is the unifying principle of biology.

   a. Molecular biology
   b. Evolution
   c. Physics
   d. Physiology
   e. Homeostasis

3. Which of the following represents a series ranked from smallest to largest?

   a. population → community → biosphere
   b. atom → molecule → cell
   c. tissue → organ → organism
   d. cell → organelle → organ
   e. a, b, and c

4. Which of the following is/are shared by all living organisms?

   a. the ability to assimilate energy
   b. the ability to transmit genetic information
   c. made up of building blocks called organs
   d. all of the above
   e. a and b only

5. Your spaceship crashes on Planet 2X-Alpha. While waiting for help, you try to discover as much as possible about your temporary home. To uncover facts about your new "natural world," you would first:

   a. Design theories to fit facts into.
   b. Do experiments.
   c. Draw conclusions.
   d. Make observations.
   e. Set up the grill and invite the residents over for dinner.

6. What's the second thing you should do to uncover facts about Planet 2X-Alpha?
   a. Design theories to fit facts into.
   b. Do experiments.
   c. Draw conclusions.
   d. Make observations.
   e. Offer to baby-sit the alien kids.

7. Which of the following is an example of a testable (falsifiable) hypothesis?
   a. Plants grow best on days when the temperature rises above 40 degrees.
   b. The best flowers are pink.
   c. Apples are nice.
   d. Many things happen, not all can be explained, but some can be experienced.
   e. Forty-two (42).

8. Your answer for the previous question was the best choice because
   a. it had the most words.
   b. it had numbers.
   c. it used inductive reasoning to generate universal laws.
   d. it was most likely to generate a controlled test.

BUT, WHAT'S IT ALL ABOUT?

Professors love to use essay questions on exams to determine if you can pull together the information you've been reading to answer a big picture question. These questions may ask you to solve a problem – practical application of knowledge. Sometimes a question will ask you to defend a model or justify a concept – construct an argument. So how do you provide the answer the professor wants? We'll show you.

At the end of each chapter, following the practice exam, we'll give you a sample essay question. In the first 10 chapters, we'll outline for you an approach to answering these "big picture" questions (beginning with chapter 11, you're on your own). What your professor wants, more often than not, is to see that you can supply specific evidence (facts and examples) that supports a more general claim made by a question. By working through these questions, you'll get some practice at organizing your information so that you can write a clear, concise answer supported by specific examples. And even if your professor never uses an essay exam, working with these questions is a good test of your understanding of the material. Let's begin.

Question: How does the fact that living organisms have a hierarchical organization impact the process of biological research?

1. **What is this question asking me to do?**
   Always stop first and figure out what the answer should look like. Is it a compare/contrast question? That means you need to provide some specific examples of similarities and differences. Does it ask you to defend a statement? Those questions require you to provide the experimental findings that support the claim made in the question. A more open-ended form of this type of question asks you to support or refute the statement. Again, you need to muster your factual evidence to defend your position. Does the question ask you to describe or document

the effect of one thing upon another, as this question does? In this case, we need to collect the information that describes the theory and describe how each part affects some other process, in this case, biological research.

2. **Collect the evidence.**

   Our first step is to define, in detail, what is meant by a "hierarchical organization" and the "process" of biological research. We start by looking for information in the textbook.

Hierarchial Organization	Process of research
Living things highly organized	Research starts with an observation that produces a question
Lower levels of organization integrated to **create** higher ones (so higher levels don't exist without lower levels)	Questions lead to proposing possible assertions that can be experimentally tested and possibly found to be false (hypotheses)
Levels = atom, molecule, organelles, cells, tissues, organs, organism, population, community, biosphere	Theory = description of general principles that fits with all of the available factual evidence. Has power to explain things but is also provisional
Levels define "building blocks"	

   Collect all the information that you can even if you probably won't use all of it in your final answer.

3. **Pull it together.**

   Since we are trying to understand the impact on the process of research, let's start by asking how each part of the research process is affected by the hierarchy of life.

   Starts with an observation

   - Can make observations on any level of the hierarchy.
   - Observations made on a given level for any one organism should be the same or very similar to that found for that same level in other organisms. For example, the properties of a specific molecule found in a fish should be very similar to the properties of the same molecule when found in a mouse.

   Questions lead to hypotheses

   - Hypotheses can be tested in different organisms as long as we consider the same level of organization – i.e., Questions about a specific molecule can be tested using molecules isolated from worms, mice, or people.
   - Testing in different organisms and collecting reproducible results in each increases our certainty in our findings.

   Evidence used to produce a theory

   - Little tricky to see how the hierarchical organizations can help us make general theories. Seems that if our research is limited to one level, the best we could do is suggest theories about one level of organization.

Did any insights occur to you as you organized this information? What about the fact that the levels of the hierarchy are integrated? Does this imply that if we define some new property of the cell that we can then suggest a new property for the tissue? Likewise, does a newly identified cellular property imply that there must be new molecules (or old molecules with previously unidentified functions) involved? We can apply what we know about the hierarchy of life to the "problem" of research method and develop the new idea that research at any one level of the hierarchy can lead to new hypotheses about function at other levels of the hierarchy.

So let's write.

The hierarchical organization of living organisms presents well-defined systems to observe and ask questions about. For example, if we are observing an organism eating, we can develop hypotheses about why it chooses that food (organism level), how easy or difficult it is to find that food (community level), how it digests that food (organ level), or how it extracts energy from that food (cellular level). So one observation can be explored on many different levels. Because all of these levels are integrated, our findings on any one level can lead us to make suggestions about function at higher or lower levels of organization. The result is that by collecting the results of research done at the molecular, cellular, tissue, organ and community levels of organization we can produce a general theory of how and why that organism chooses that particular food.

2

CHAPTER 2

The Fundamental Building Blocks: Chemistry and Life

Basic Concepts

- The fundamental building block of matter is the atom, an elemental particle composed of protons, neutrons, and electrons.
- The complex molecules of life are created from atoms through covalent, ionic, and hydrogen bonding.
- Chemical bonds define a molecule's physical, chemical, and biological properties.

Fundamental Building Blocks of Matter

- The basic forms of matter are the elements, 92 pure substances that cannot be reduced to simpler components. A single, elemental particle is called an atom.
- Atoms contain protons, neutrons, and electrons. Protons and neutrons occupy the nucleus of an atom and account for its mass. Electrons are essentially massless particles that occupy the space around the nucleus.
- Atoms are electrically neutral because, although protons are positively charged and electrons are negatively charged, the number of protons equals the number of electrons, assuring no net charge.

THINK ABOUT IT

1. **Create a web** that defines matter using the terms introduced in section 2.1.

2. Fill in the following table with the three subatomic particles and their properties.

Particle Name	Location in Atom	Charge	Mass Contribution

3. Describe the difference between atomic number and atomic weight.

4. Which of the following atoms would be isotopes of the same element?

   a. Atomic number = 6 Atomic weight = 12
   b. Atomic number = 1 Atomic weight = 2
   c. Atomic number = 6 Atomic weight = 13
   d. Atomic number = 8 Atomic weight = 16
   e. Atomic number = 6 Atomic weight = 14
   f. Atomic number = 1 Atomic weight = 14

Chemical Bonding

- We think of electrons surrounding the nucleus in layers, called shells, that reflect the energy content of the molecule. A limited number of electrons can occupy any given shell and, when a shell is filled, the atom exists in its most stable form.
- Covalent chemical bonding creates filled electron shells because the atoms involved are able to share electrons to achieve a full, outer valence shell. The total number of electrons does not change when a covalent bond is made. A molecule is created when two or more elements are joined by covalent bonds.
- Electrons in a covalent bond may not be shared equally. Unequal sharing produces a small difference in the charge distribution around the molecule, so that one end is more positive and the other is more negative, creating a polarity to the bond. Polar molecules may associate through very weak bonds called hydrogen bonds.
- Ionic chemical bonding occurs when electrons are pulled from the shell of one atom into the shell of another, creating two charged species known as ions. The atom that has lost an electron becomes a positive ion; the one that gains an electron becomes a negative ion. The ions remain associated because their opposite charges attract each other.

THINK ABOUT IT

1. a. Which subatomic particle(s) determine the identity of an element?

   b. Which subatomic particle(s) determine the interactions between atoms?

2. Using this figure from your text, describe why water is a polar molecule. What does the polarity of a water molecule imply about the behavior of water molecules?

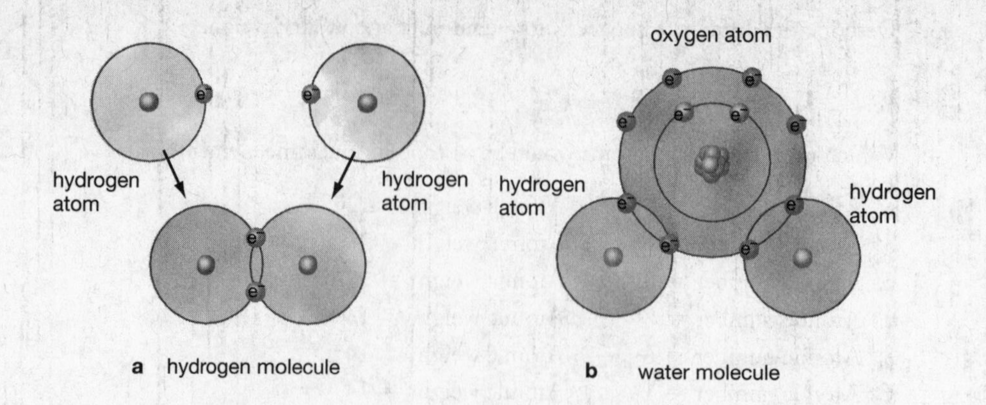

3. Fill out the following table describing the three types of bonds described in this section.

Bond	Atom Interactions

4. Which of the following atoms would be considered an ion?

   a. a sodium atom with 11 protons, 11 neutrons, 11 electrons
   b. a chlorine atom with 17 protons, 17 neutrons, 18 electrons
   c. a magnesium atom with 12 protons, 12 neutrons, 10 electrons
   d. a sodium atom with 11 protons, 11 neutrons, 10 electrons

Some Properties of Chemical Compounds

- Molecules have a three-dimensional shape that determines its own chemical and biologic activity and its ability to interact with other molecules.
- Shape and polarity determines how well molecules form homogeneous mixtures called solutions. Solvent molecules break the chemical bonds, usually ionic or hydrogen bonds, that hold solute molecules together.
- "Like" dissolves "like"—molecules of similar polarity can form solutions, like water and salt, that are both polar molecules.

THINK ABOUT IT

1. Why is the shape of a biological molecule important?

2. You like your coffee black with lots of sugar. (a) What is the solvent? (b) What is the solute? (c) What kind of chemical bonds are broken in the solute to make a solution?

3. Give three reasons why water is important to living organisms.

Cool Fact: X-ray Crystallography

The shape of a biological molecule determines its activity, yet how can we see the shape of a molecule? Molecules are generally too small to see in a microscope—even in an electron microscope, which uses beams of electrons to create a picture of very small things. Fortunately, we can take advantage of two properties of matter and a great deal of computer technology to create pictures of complex biomolecules.

Matter, as we know, has all of its mass concentrated in the nucleus of the atoms that compose the molecules. When we shine on an object such as a book, it reflects the light back to us. Because the light cannot pass through the book, we see a book. Light waves are too long to produce a reflection of a nucleus, but x-rays are much, much shorter. We can bounce x-rays off the nucleus of an atom, but because we cannot see x-rays, we use some other detector (like film) to collect the reflected pattern. Using a mathematical technique called a Fourier transform and lots of computer power, we can translate the reflected pattern into a picture showing the positions, in space, of these nuclei relative to each other.

Pictures of single atoms are not very useful, however. But, chemical bonds hold nuclei together, and the type of bond and the atoms involved defines this distance. So, when we make an x-ray picture of a molecule, we learn how these nuclei are arranged in space relative to each other. Again, improvements in computer technology have allowed scientists to interpret the complicated reflection patterns produced by bouncing x-rays off the atoms of very large molecules so that pictures can be "drawn" of many important biological molecules.

Biological molecules, however, occur in aqueous solutions, so they rarely hold still "for the camera." The art of this picture-taking technique, called x-ray crystallography, is making these molecules stay still by forcing them to form crystals. Defining the environmental conditions that will force large molecules to form a crystal is still a process of trial and error. But scientists will go to great lengths to optimize their crystal-growing conditions, including sending biomolecular solutions into space on the Space Shuttle to take advantage of zero-gravity conditions. In return, humans gain the rare opportunity to meet, face to face, the molecules that make life possible.

Chapter 2

TESTING, TESTING

Vocabulary Review

Place each term from the following list into the appropriate column of the table below. Some terms may belong in more than one column.

chemistry	isotope	proton
covalent	neutron	solute
electron	nonpolar	solvent
ion	nucleus	valence shell
ionic	polar	

Matters	Bonds	Solutions

PRACTICE TEST

Once you have finished studying this chapter, close your books, grab a pencil, and spend the next 15 to 20 minutes completing this practice test.

Compare and Contrast

For each of the following paired terms, use a sentence of comparison ("Both . . . ") and a sentence of contrast ("However, . . . ").

polar/nonpolar

inert/reactive

covalent/ionic

element/matter

Matching

Match the following terms with their description. Each choice may be used once, more than once, or not at all.

a. protons
b. neutrons
c. electrons
d. ions
e. isotopes

__ Found in the nucleus, has no charge
__ Has gained or lost electrons and will form bonds because of this
__ Found in the nucleus, has charge and mass
__ Has charge, but no mass
__ Differs in the number of neutrons from other atoms of its type

Short Answer

1. Why are there different types of bonds?

2. What kinds of bonds would you expect nonreactive or inert elements to form?

3. Which would have fewer electrons, an atom or its negatively charged ion?

4. An atom of an element has 6 electrons in its outermost shell. How many does it have in its innermost shell?

Multiple Choice

1. Nitrogen has 7 electrons. How many covalent bonds will it form?

   a. 7
   b. 3
   c. 5
   d. 10
   e. 12

2. You discover a new element, Mimionium, that has 400 protons and 401 neutrons. You predict that the atomic form of Mimionium will have ________ electrons and that the negatively charged ion ($Mimi^-$) will have _________ electrons.

   a. 801, 401
   b. 401, 399
   c. 400, 401
   d. 401, 800
   e. 400, 399

3. Hydrogen bonding between water molecules occurs between:

   a. the partially positive H of one molecule and a partially negative O of another.
   b. the oxygen atom of one molecule and the hydrogen atom in the same molecule.
   c. the partially negative H of one molecule and a partially positive O of another.
   d. the partially negative H of one molecule and the partially positive H of another.
   e. the partially negative O of one molecule and the partially negative O of another.

4. Which of the following atoms would you expect to be extremely reactive?

   a. Ne, atomic number = 10
   b. C, atomic number = 4
   c. Ar, atomic number = 18
   d. Na, atomic number = 11
   e. O, atomic number = 8

5. Which of the following best describes an ion?

   a. an atom in which the numbers of protons and neutrons differ
   b. an atom in which the numbers of protons and electrons differ
   c. an atom in which the numbers of neutrons and electrons differ
   d. an atom in which the number of protons equals the number of electrons
   e. an atom in which the number of protons equals the number of neutrons

6. Which of the following best describes the "rules" for solutions?

   a. Opposites attract.
   b. Generalizations are not useful.
   c. Only solids can be solutes.
   d. Like dissolves like.
   e. Only solids can be solvents.

7. The outermost layer of the atom, where electrons are gained, lost, or shared is the

   a. seashell
   b. electron shell
   c. valence shell
   d. taco shell
   e. orbital shell

8. Rank the following terms from least to most inclusive.

   a. neutron, electron, atom, element, matter, solution
   b. matter, electron, atom, neutron, element, solution
   c. electron, neutron, atom, element, solution, matter
   d. atom, electron, neutron, matter, element, solution
   e. element, atom, matter, electron, neutron, solution

9. What is the bond that forms between water molecules called?

   a. ionic
   b. polar covalent
   c. nonpolar covalent
   d. hydrogen
   e. isotopic

10. Molecule A is made up of carbon, hydrogen, and nitrogen atoms. Molecule B contains the same elements in the same proportions, but has a different three-dimensional shape. What other differences would you expect to find between the two molecules?

    a. Their protons would have different shapes.
    b. Their elements would be different.
    c. Their properties would be different.
    d. Their electrons would be of different types.
    e. They would drive different cars.

11. Free radicals have been blamed for scarring the walls of blood vessels and mutational damage to the DNA found in mitochondria. These radicals are formed in response to oxygen's attraction for

    a. protons
    b. ions
    c. electrons
    d. neutrons
    e. hydrogen bonds

12. Had medieval alchemists been successful in changing one element into another what would they actually have changed about the atoms of the first element? In other words, what truly defines an element?

    a. the number of electron shells
    b. the number of covalent bonds it can form
    c. the number of neutrons
    d. the number of electrons
    e. the number of protons

13. Your new understanding of chemistry has lead you to take a job as a research chemist. You discover four new elements with the following properties. Which of these is *least* likely to form bonds with other elements?

    a. element Kc has 4 electrons in its outermost shell
    b. element Kd has 8 electrons in its outermost shell
    c. element Sb has 3 electrons in its outermost shell
    d. element Lf has 6 electrons in its outermost shell
    e. all of these elements are likely to form bonds with other atoms.

BUT, WHAT'S IT ALL ABOUT?

Here's a question to help you pull together what you've learned so far using this text. If you don't remember how to attack this type of question, check out the example in Chapter 1.

Question: How does the type of bonding in a molecule affect its properties?

1. **What kind of question is this?**
   This question wants to know one "thing," in this case bonding, impacts another thing, molecular properties.

2. **Collect the evidence**
   What do we need to know to answer this question?

   a. Types and properties of molecular bonds
   b. Role of bonds in the properties of biological molecules

3. **Pull it all together**
   Provide an example of each type of bond and describe how its properties work to hold molecules together. You can use the following topic sentences to get you started.

   Bonds create stable structures.

   The solubility of a molecule is a function of its type of bonds.

   Another property affected by bonding is the three-dimensional shape of a molecule.

3

CHAPTER 3

Water, pH, and Biological Molecules

Basic Concepts

- Water's unique structure and polar character facilitate its role as a biologic solvent.
- We measure the dissociation of water, and other molecules, using the pH scale. Solutions with a low pH are acids; those with a high pH are bases.
- All the important molecules of life—carbohydrates, proteins, lipids, and nucleic acids—use the element carbon as a starting point.
- Biomolecules are built from monomeric units—carbohydrates from simple sugars, proteins from amino acids, and nucleic acids from nucleotides.

Water

- Water works as a solvent because it can form hydrogen bonds with many molecules. It can also form a hydrogen bond with itself, which explains its ability to serve as a heat buffer, to become less dense in its solid state, and to have a high surface tension.
- Water cannot solubilize nonpolar molecules, like hydrocarbons, because it cannot disrupt the attraction that holds one nonpolar molecule to another. Nonpolar molecules are referred to as hydrophobic because they do not interact with water, like hydrophilic molecules do.
- Water's ability to disrupt the hydrogen bonds formed when hydrophilic molecules interact creates acids and bases. The pH scale allows us to quantify how readily a molecule will lose a hydrogen ion (acidic) or gain a hydrogen ion (basic) when put in an aqueous solution.
- The pH scale is logarithmic—a solution at pH 5 has 100 times more hydrogen ions ($H+$) in it than a solution at pH 7. Smaller numbers mean more hydrogen ions are present and the solution is more acidic; bigger numbers mean the solution is less acidic and more basic.

THINK ABOUT IT

1. Explain why the following properties of water are important to life.

   a. hydrogen bond formation

   b. density

c. specific heat

d. cohesion

2. Your home is in Boston, Massachusetts, and your college roommate is from Tucson, Arizona. Given that Boston is near the ocean and Tucson is not near any bodies of water, how would you expect daily and nightly temperatures to compare in these two cities?

3. Your pet iguana loves sunlight. Would you say that your iguana is photophobic or photophilic?

4. You have two chemicals. You dissolve each in water and make the following data table based on your observations. Fill in the missing information.

	pH	Acid or Base?	Relative Strength
Chemical 1	4		
Chemical 2	2		

5. Draw your own pH scale.

—Where would vinegar (acetic acid) be?
—Where would you place Agent X, that decreases the amount of H+ in solution? (neutral, less than pH 7, or greater than pH 7)
—How does the concentration of H+ change if the pH value drops from 4 to 3? From 4 to 2?

Carbon as the Building Block of Biomolecules

- One carbon atom can make four covalent bonds. It can bond to itself as readily as to oxygen and hydrogen, and can make molecules with a variety of shapes. Living things may be thought of as solutions of carbon-containing molecules.
- Carbon-based biomolecules fall into four large categories—carbohydrates, lipids, proteins, and nucleic acids.

THINK ABOUT IT

1. Carbon has four electrons in its valence shell. What types of bonds would you expect it to form, and how many per atom of carbon?

Polymers from Monomers: Carbohydrates, Lipids, Proteins, and Nucleic Acids

- Carbohydrate monomers, called monosaccharides, are composed of carbon, hydrogen, and oxygen. These monomers form polymers, called polysaccharides, that can be used as food (such as starch), to store energy (such as glycogen), or to give structure (such as chitin and cellulose).
- Lipids are a diverse collection of molecules that share the property of being hydrophobic. Some, like triglycerides, serve as a source and storage form of energy, whereas others, like steroids and phospholipids, serve a structural function. The monomeric unit of all lipids, except the steroids, is the fatty acid— a long chain of carbon and hydrogen atoms ending in a carboxylic acid.
- Proteins are polymers made from any combination of 20 monomeric units called amino acids (carbon-based molecules that also contain a nitrogen atom). Although proteins are linear polymers, they can fold to assume a variety of shapes or conformations. The conformation of a protein is critical to its function.
- Nucleic acids serve a variety of functions, but are most important for their ability to store and transmit information for the construction of protein polymers. Nucleic acids, DNA and RNA, are made of nucleotide monomers that contain a phosphate group, a monosaccharide, and a nitrogen-containing base.

THINK ABOUT IT

Section 3.4 of the chapter describes in detail the four classes of biomolecules, beginning with the structures of the monomeric units used to make each biomolecule, showing you how the polymeric molecules are assembled, and describing the unique properties of each type of biomolecule. Using the basic skeleton that follows, create your own outline of this section of Chapter 3.

Biomolecules

Monomeric units – List names, chemical composition, features of their structure.

Polymeric units – How are they assembled? List examples of each type of polymer. What function do they serve?

Unique properties of the class – Do the monomers and/or polymers of this class of molecules perform special functions for the organism? Do they have unique physical or chemical properties (shape, charge, ability to be dissolved by water)?

1. Use the following table to name the building blocks and some functions of the following macromolecules.

Molecule	Building Blocks	Function 1	Function 2
carbohydrates			
glycerides			
proteins			
nucleic acids			

2. For the following carbohydrates, state whether they are simple or complex, and briefly describe their function.

Carbohydrate	Complex or Simple	Function
glucose		
cellulose		
chitin		

3. What are two major functions of lipids?

4. a. What is the difference between a saturated and unsaturated fatty acid?

   b. Which one is likely to be solid at room temperature?

5. a. Draw a triglyceride and label the following: glycerol, fatty acids.

   b. Where does the condensation reaction occur on this molecule?

6. a. The steroids share a common skeleton. What is it?

   b. What is unique to each steroid?

   c. Name two steroids.

7. Why is the shape of a protein so important?

8. Name the molecule illustrated here. Name the functional groups.

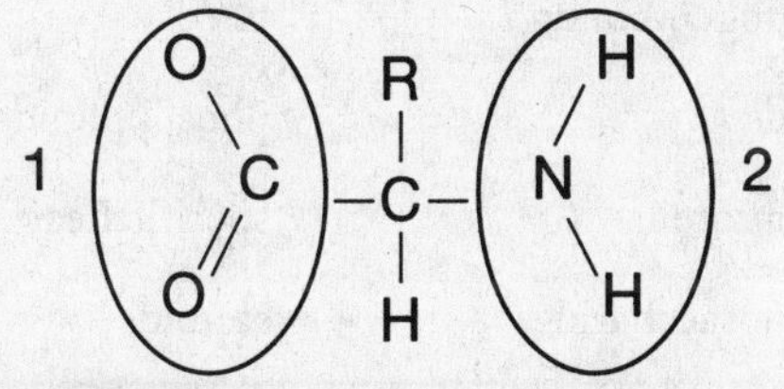

   What is the function of the R group?

9. What do nucleic acids do?

10. What are the three parts of a nucleotide?

Cool Fact: Frozen Frogs

Ice, solid water, floats. This unique property of water has significant ecological consequences. In the winter, bodies of water freeze from the top down so that aquatic animals can survive beneath a layer of insulating ice all winter. But what about animals that don't live in water? How do they deal with freezing temperatures?

Animals have a variety of ways to protect themselves from falling temperatures, some more unusual than others. For example, some frogs actually freeze and stay in a state of suspended animation over the winter. To prepare for freezing, the frogs increase the solute concentration in their cells by making and storing more carbohydrate molecules. These sugars act as antifreeze, lowering the freezing point for the cytoplasm of the frog's cells. The fluid surrounding the cells freezes solid, but the cells themselves are protected from the expanding ice crystals. Come spring, the extracellular fluid thaws, the frog's cells are intact, and the animal gets back to business.

Some animals that hibernate adopt a similar strategy. Arctic ground squirrels don't freeze solid like the frogs, but they can drop their body temperature to below freezing. They prevent ice damage within the cells by eliminating molecules that could nucleate an ice crystal. Spending the winter in a supercool state eliminates this animal's need to use energy to maintain a high body temperature.

Chapter 3

TESTING, TESTING

Create a web that illustrates the relationship between water, carbohydrates, carbon, nitrogen, proteins, lipids, oxygen, hydrogen, and nucleic acids.

Vocabulary Review

Place each term from the following list into the appropriate column of the table below. Some terms may belong in more than one column.

acid	fats	oils
base	hydrogen bonds	pH
buffer	lipids	phospholipids
cohesion	lipoprotein	specific heat
DNA	monosaccharides	steroids

Water	Acid/Base	Molecules

PRACTICE TEST

Once you have finished studying this chapter, close your books, grab a pencil, and spend the next 15 to 20 minutes completing this practice test.

Compare and Contrast

For each of the following paired terms, write a sentence of comparison ("Both . . . ") and a sentence of contrast ("However, . . . ").

hydrophobic/hydrophilic

DNA/RNA

monosaccharide/polysaccharide

lipoprotein/glycoprotein

Matching

Match the following terms with their description. Each choice may be used once, more than once, or not at all.

a. chitin
b. cellulose
c. starch
d. glycogen
e. hydrocarbons

___ Energy storage in animals
___ Energy storage in plants
___ Structural sugar in roach shells
___ Source of insoluble fiber in the diet
___ Made of hydrogen, oxygen, and carbon
___ Organic, but not an important molecule in living cells

Short Answer

1. What property of water creates surface tension?

2. Why does adding a base to water make it more basic?

3. Why do we describe life on Earth as being carbon-based?

4. Why is it preferable to have a diet low in saturated fats?

Multiple Choice

Circle the letter that best answers the question.

1. Why does it take a lot of energy to raise the temperature of water compared to other solvents?

   a. because liquid water is "locked" into a rigid lattice of molecules that doesn't break easily
   b. because water has a low specific heat
   c. because energy is needed to break the hydrogen bonds between water molecules

d. because water must be stored in containers which resist the transfer of heat
e. none of the above

2. Why do living beings have buffers?

   a. to maintain an acidic pH
   b. to carry out chemical reactions
   c. to maintain a pH close to neutral
   d. to maintain a basic pH
   e. b and d

3. Which of the following macromolecules is/are not technically polymeric?

   a. glucose
   b. glycogen
   c. DNA
   d. steroids
   e. a and d

4. Which of the following protein structures describes the association of two or more proteins?

   a. primary structure
   b. quaternary structure
   c. denaturation
   d. tertiary structure
   e. secondary structure

5. Hydrangeas are a popular type of garden plant. Certain varieties produce flowers whose color is dependent on the pH of the soil in which the plant is grown. Soil below a pH of 7 produces blue flowers; soil above 7 produces pink flowers. Which of the following substances would you add to the soil to produce pink flowers?

   a. a base
   b. an acid
   c. a buffer
   d. either b or c
   e. a, b, or c

6. RNA is a polymer of:

   a. amino acids
   b. glucose
   c. phospholipids
   d. steroids
   e. nucleotides

7. Let's suppose that by some strange circumstance (the conjunction of the planets, alien abduction, shifts in the Earth's magnetic core, etc.) your best friend, Pat, becomes allergic to hydrogen ions (H+). Which of the following would have the lowest concentration of hydrogen ions and therefore be safest to give Pat?

   a. Coffee, pH 5
   b. Lemon juice, pH 2
   c. Water, pH 7
   d. Cola, pH 3
   e. Urine, pH 6

8. Which of the following is not used as an energy source?

   a. fats
   b. starches
   c. glycogen
   d. oils
   e. nucleic acids

9. Which of the following is a product of a condensation reaction?

   a. ions
   b. proteins
   c. molecule of H_2O
   d. atoms
   e. canned soup

10. What is the function of phospholipids?

    a. forms insect shells
    b. forms plant fibers
    c. forms enzymes
    d. forms cell membranes
    e. makes up genetic material

11. One way to help cure world hunger would be to find an enzyme that would allow humans to digest cellulose. This would work because cellulose is a type of __________, which serves as a source of energy in our bodies.

    a. protein
    b. carbohydrate
    c. lipid
    d. nucleic acid
    e. glue

12. Sickle-cell anemia is a disease that is caused by the change in shape of the hemoglobin molecule. Because hemoglobin is a protein, you know that the mutation

that causes this change must occur in the ________ structure of the protein because it is at this level that three-dimensional shape is determined.

a. primary
b. secondary
c. tertiary
d. quaternary
e. dictionary

13. If water were more dense at 4 degrees Celsius instead of LESS dense then . . .

a. ice cubes wouldn't float in your lemonade.
b. aquatic vertebrates would have trouble surviving the winter in northern areas.
c. water in deep ponds and lakes would freeze solid every winter.
d. All of these would happen.
e. None of these would happen.

14. You happen to be out shopping when you spy your favorite movie star. Excitement causes you to begin hyperventilating which causes your blood to

a. become more basic
b. become more acidic
c. become neutral
d. your blood pH will not change

15. After your yearly physical, you vow to pay more attention to your health. A quick run to the grocery store fills your shopping cart with the following items. Which one is the best choice for lowering your blood cholesterol levels?

a. a spread comprised of monounsaturated olive oil
b. butter
c. an omega-3 fatty acid salad dressing
d. an artificial margarine made from trans fatty acids

16. Suppose you have discovered a wonder drug that would cure most of the common human illnesses. Unfortunately the molecule is hydrophobic and will not dissolve into a solution that can be administered to patients. Which functional group should you add to your molecule to make it more like ethanol, which IS water soluble?

a. hydroxyl ($-OH$)
b. amino ($-NH_2$)
c. carboxyl ($-COOH$)
d. phosphate ($-PO_4$)

BUT, WHAT'S IT ALL ABOUT?

Here's a question to help you pull together what you've learned so far using this text. If you don't remember how to attack this type of question, check out the example in Chapter 1.

Question: Most biologically important molecules are polymers built from monomeric units. Why is this advantageous to a living organism? Support with examples.

1. **What kind of question is this?**
   This is a variation on the "defend a statement" question. You need to provide evidence to support the claim that making large molecules out of "building blocks" provides some advantage to the organism.

2. **Collect the evidence.**
   "Advantages" will be general properties of molecules that serve a purpose. To document that these are advantages, you need to find specific examples of how polymer can be more functional than its monomeric unit.

3. **Pull it all together.**
   Each advantage that you identify would be the topic of a paragraph. The body of each paragraph would provide specific examples of how building polymers from monomers produces the advantage you claim. You can use the following topic sentences to get you started.

   Creating polymers from monomer offers two advantages to the organism: ease of molecular construction and increased diversity of molecules.

   In addition to easy construction, the cell can make many different molecules from the same collection of monomers just by changing the order of the monomers.

4

CHAPTER 4

Life's Home: The Cell

Basic Concepts

- All cells are either prokaryotic, like bacteria and archaea, or eukaryotic, like protista, fungi, plants, and animals.
- The two types of eukaryotic cells—animal cells and plant cells—possess a nucleus and other intracellular organelles that allow for compartmentalization of cellular activities.
- Plant cells differ from animal cells because they have (1) cell walls, (2) a central vacuole for water storage, and (3) plastids such as chloroplasts.

Prokaryotic versus Eukaryotic

- All living organisms are composed of cells that share a fundamental unity of structure. All cells have at least one membrane that separates the contents of the cell from its surroundings. Cells contain specialized molecules that permit the reactions of life to occur. All cells can make copies of themselves.
- The prototypical prokaryotic cell is a bacteria—a single-cell organism that lacks intracellular organelles. Prokaryotic cells are smaller and functionally simpler than eukaryotic cells.
- Compartmentalization of specialized functions within membrane-bound organelles of eukaryotic cells facilitates formation of tissues and organs. Both plants and animals are comprised of eukaryotic cells. Plant cells differ from animal cells somewhat in their complement of organelles, but both types of eukaryotic cells share many structures and processes.

THINK ABOUT IT

1. Label the following with a "P" for prokaryote or "E" for eukaryote.

___dog	___bacteria	___bird	___tree
___no nucleus	___multicellular	___true nucleus	___fungus
___cell wall	___organelles	___central vacuole	___mitochondria

2. Give one specific example of a compartment within a eukaryotic cell and briefly describe its function. Do prokaryotic cells have the ability to perform this same function?

Tour of the Eukaryotic Animal Cell

- The nucleus contains the genetic material (DNA) that controls the structure and reproduction of the cell. The nucleus also contains the nucleolus, which makes the RNA molecules that will become the ribosomes.
- Organelles—such as mitochondria, endoplasmic reticulum, and lysosomes—are membrane-delimited structures that allow for cellular activities to be isolated from each other. Mitochondria produce energy for the cell, whereas the acidic lysosomes break down and recycle cellular molecules; membranes prevent the work of one set of reactions from interfering with another set of reactions.
- Organelles are suspended in a protein-rich, aqueous environment called the cytosol. Cells are not just sacks of structures; cells have architecture provided by the internal scaffolding of the cytoskeleton. The cytoskeleton gives animal cells a flexible shape.
- Proteins move around within the cell, and from the inside to the outside of the cell, through the action of an endomembrane system comprised of the rough endoplasmic reticulum (RER) and the Golgi. Proteins made in the RER are packaged into vesicles that can move through the cytoplasm to an appropriate destination, or that can fuse with the plasma membrane and release their contents in the process of exocytosis.

THINK ABOUT IT

1. a. Arrange the following terms in the order that represents the correct flow of information:

protein	ribosome	nucleus	mRNA	cytoplasm
DNA	nuclear pore	vesicles	nuclear pore	vesicles

   b. Where does the process of exocytosis fit into the flow of information in question 1a?

2. You visit a factory that makes jam and then ships bottles of jam to various grocery stores to be sold. Which cellular organelle would correspond to the jam production line? Which organelle would correspond to the factory's shipping department?

3. a. What is the structural difference between rough ER and smooth ER?

b. What are two functions of the smooth ER?

(i)

(ii)

4. Why is it necessary to package digestive enzymes within the lysosome?

5. Which organelle provides the energy you use to read this sentence?

6. Draw a flowchart that illustrates the movement of proteins to the surface of the cell.

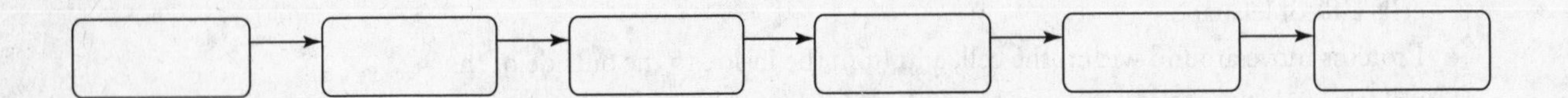

Tour of the Eukaryotic Plant Cell

- The primary feature that distinguishes a plant cell from an animal cell is the cell wall, a thick, rigid structure composed mostly of cellulose. The cell wall gives plants their distinctive shape and also limits water movement into and out of the cell.
- The central vacuole dominates the cytoplasm of the plant cell. This vacuole is a primary site for metabolism because it stores nutrients, processes waste products, stores pigment molecules, and controls the internal pH of the cell.
- Plant cells contain organelles called plastids. Plastids also store pigments and process nutrients. The best-known plastids are the chloroplasts, the site of photosynthesis in green plants and algae.

THINK ABOUT IT

1. Name three specialized structures of plant cells that are not present in animal cells.

2. a. Which plant-cell organelle produces oxygen?

   b. Which cell organelle consumes oxygen?

No Eukaryotic Cell Needs to Be an Island: Cell-Cell Communication

- Multicellularity is no advantage if each cell still functions as a separate entity. Cells within tissues and organs communicate with each other through channels. Plant-cell channels are called plasmodesmata and must penetrate the cell wall. Animal-cell channels are called gap junctions.
- Plasmodesmata and gap junctions are specialized protein structures that control the flow of molecules and electrical signals between cells.

THINK ABOUT IT

1. Why do you think plasmodesmata are permanent while gap junctions are temporary? What is the major difference between the outer boundaries of plant and animal cells?

2. What molecule forms the channels between these cells?

Cool Fact: Cellular Communication and Cancer

The disease cancer is thought to be a multistep process. First, something needs to stably alter the DNA in a cell; that is, a mutation must occur. Mutations can be the result of a mistake made during the replication of DNA during the normal cell division process, or they can be caused by the presence of an external agent (such as UV light or various chemicals) that damages DNA. If the molecules made by this damaged DNA can no longer control cell replication or growth, then the cell could begin to replicate more often than it should, producing a tumor.

A mutation alone does not produce a tumor. This mutated cell—and cancer is believed to develop from a single cell—needs another stimulus that causes the expression of this damaged DNA. Many chemicals are known to stimulate messenger RNA synthesis or protein translation, so the presence of one of these chemicals would trigger the next stage in tumor development—production of a mutated protein and loss of control of cellular replication.

Mutated cells can be kept under control by their normal neighbors via the chemical signals that pass through the gap junctions. But if the gap junctions are damaged, cell-cell communication stops and the mutated cell loses its "don't replicate" signals. Mice that lack the major gap-junction protein, connexin, are much more likely to develop tumors in response to DNA damage. Likewise, chemicals that promote tumor development, like PCP, have been shown to inhibit gap junction function in vitro. Drugs that restore gap-junction function, then, could reverse or even prevent the development of cancer.

TESTING, TESTING

Create a web that demonstrates the relationship between nucleus, plant cell, animal cell, mitochondria, membrane, prokaryote, and plastid.

Vocabulary Review

Place each term from the following list into the appropriate column of the table below. Some terms may belong in more than one column.

cell wall
gap junctions
microtubule
nucleus
plasmodesma
central vacuole
Golgi
mitochondria
plasma membrane
rough ER
chloroplast
lysosome
nucleolus

ANIMAL CELL	PLANT CELL

PRACTICE TEST

Once you have finished studying this chapter, close your books, grab a pencil, and spend the next 15 to 20 minutes completing this practice test.

Compare and Contrast

For each of the following paired terms, use a sentence of comparison ("Both . . . ") and a sentence of contrast ("However, . . . ").

cilia/flagella

nucleus/nucleolus

cytosol/cytoplasm

smooth ER/rough ER

Matching

Match the following terms with their description. Each choice may be used once, more than once, or not at all.

a. lysosome
b. nucleus
c. Golgi
d. plastid
e. transport vesicle

___ Sorts and ships secretory proteins
___ Stores material in plant cells
___ Command center of cell
___ Recycling center of cell
___ Moves material from one location to another

Short Answer

1. Plants have plasmodesmata and animals have gap junctions for communication. Why is this ability important for multicellular organisms?

2. Why should we not be surprised that plant cells have central vacuoles and walls and animal cells don't?

3. If the cell is a factory, churning out materials, what factory part or function is analogous to the mitochondria?

4. What would happen to the proteins synthesized on the rough ER if the Golgi complex were not present?

Mupltiple Choice

Circle the letter that best answers the question.

1. The smallest of the cytoskeleton elements is the

   a. microtubule.
   b. intermediate filament.
   c. micromini.
   d. microfilament.
   e. none of the above.

2. The largest structure visible within an animal cell is usually the

   a. mitochondria.
   b. Golgi apparatus.
   c. ribosomes.
   d. nucleus.
   e. cytosol.

3. Which is the following is the building site for proteins?

   a. nucleolus
   b. smooth endoplasmic reticulum
   c. ribosome
   d. DNA
   e. shop class 101

4. Which of the following is a site for lipid synthesis?

   a. lysosomes
   b. rough endoplasmic reticulum
   c. smooth endoplasmic reticulum
   d. nucleus
   e. plasma membrane

5. Which organelle is the recycling center of the cell?

   a. mitochondria
   b. the nucleus
   c. smooth ER
   d. the lysosome
   e. the Golgi body

6. What types of eukaryotic cells are flagellated?

   a. motile bacteria
   b. all eukaryotic cells
   c. sperm
   d. no eukaryotic cells
   e. jellyfish cells

7. Which of the following eukaryotic organelles is (are) specific to plant cells?

   a. chloroplast
   b. cell wall
   c. mitochondria
   d. all of the above
   e. a and b only

8. Which of the following structures permit(s) communication between eukaryotic cells?

   a. gap junctions
   b. plasmodesmata
   c. central vacuole
   d. microfilaments
   e. a and b

After collecting samples of the material found growing on last month's leftovers at the back of your fridge, you realize that you have discovered a new type of organism. You find that the cells have distinct subunits within. Further, several of these subunits, which you have nicknamed Groucho, Chico, and Harpo, seem to be involved in the production and transfer of certain molecules. Molecule Dumont, as you call it, appears after a messenger molecule is sent from Groucho to Chico. Dumont then travels to Harpo, where it is covered in membrane and (eventually) released from the cell.

9. What type of cells do you think these are?

   a. prokaryote
   b. eukaryote
   c. neither

10. Dumont is most likely what type of molecule?

    a. protein
    b. lipid
    c. carbohydrate
    d. DNA
    e. water

11. Which of the following is the likely nucleus of these cells?

    a. Harpo
    b. Groucho
    c. Chico
    d. Dumont
    e. Zeppo

12. Which of the following has the same function as the Golgi complex?

    a. Harpo
    b. Groucho
    c. Chico
    d. Dumont
    e. Zeppo

BUT, WHAT'S IT ALL ABOUT?

Here's a question to help you pull together what you've learned so far using this text. If you don't remember how to attack this type of question, check out the example in Chapter 1.

In this chapter, we've learned that all living things are cells or are made of cells. Because we know that there are four classes of biomolecules (proteins, lipids, carbohydrates, and nucleic acids), then it follows that cells and the structures within the cells must be made from these biomolecules. How do these molecules shape the structure and the function of cellular components?

1. **What kind of question is this?**
   This question wants to know one "thing," in this case a type of molecule, impacts two other things, higher order cellular structure and the functioning of those structures.

2. **Collect the evidence.**
   What do we need to know to answer this question?
   a. Molecular composition of cellular structures
   b. Function of cellular structures

3. **Pull it all together.**
   For each class of biomolecule, provide an example of a type of molecule that makes up a particular cellular structure and describe why this type of molecule is important to the "job" of that structure. You can use these first two sentences to get you started.

   All parts of the cell are made from the four basic types of molecules: lipids, proteins, nucleic acids, and carbohydrates. Each class of biomolecules contributes to the specialized functions of the cells component parts.

CHAPTER 5

Life's Border: The Plasma Membrane

Basic Concepts

- The functionality of the plasma membrane relies upon its structure.
- Water moves freely across the membrane, down a concentration gradient, in a process called osmosis. Some small molecules also freely diffuse through the cell membrane.
- Transport of most molecules across the membrane requires specialized membrane structures. The process can be either passive, down a concentration gradient, or active, against a concentration gradient.
- Very large molecules must be packaged to move into or out of the cell. These processes are called endocytosis and exocytosis, respectively.

Plasma Membrane—Structure and Function

- The plasma membrane functions first as a barrier—it keeps biologic molecules concentrated inside the cell and keeps harmful substances outside.
- Membranes are made of a bilayer of phospholipids. The phospholipids are oriented such that their hydrophobic heads face the aqueous environment (on the inside and outside of the cell) whereas their hydrophobic tails face each other.
- Other lipids molecules, like cholesterol, help control the fluidity of the membrane.
- Proteins within the lipid bilayer allow selective movement of molecules into and out of the cell. Proteins may also interact with cytosolic molecules and help determine the shape of the membrane.
- Various molecules embedded in the membrane receive information from signal molecules outside of the cell.

THINK ABOUT IT

1. What are the four functions of cell membranes?

2. Which function is defective in a cystic fibrosis patient like Gunner Esiason?

3. What functions do proteins carry out on the external surface of the cytoplasmic membrane?

4. a. What part of the cytoplasmic membrane most closely resembles icing on a cake?

   b. What are the constituents of this structure, and what do they do?

5. Compare and contrast phospholipids and cholesterol. Which one makes up the bulk of the cell membrane?

Diffusion and Osmosis

- Diffusion occurs as molecules randomly move from an area of high concentration to one of low concentration. When the molecules moving are water, we call the process osmosis.
- The cell membrane is a semipermeable barrier; water can move across it freely despite its polarity, but most solutes cannot. This selective movement of solutes creates a concentration gradient between the inside and outside of the cell.
- When the movement of solutes is limited, water moves through the cell membrane (osmosis) to balance the solute concentration inside and outside of the cell.

THINK ABOUT IT

1. If shipwrecked sailors die after drinking lots of highly concentrated seawater, what do you predict would happen to the cells of a person after drinking copious quantities of distilled water? (Distilled water has no dissolved solutes of any kind.)

2. How does the accumulation of chloride ions in Gunnar Esiason's lung cells lead to a thickening and drying of the mucus in his lungs?

3. Which of the following substances would you expect to cross the membrane "solo"? For each, explain your answer.

Substance	Cross Solo?	Reason
Water		
Oxygen		
A large protein		
A small molecule with a large charge		
A small molecule with a large charge		

4. On the diagram below, use an arrow to show the movement of water. The patterns symbolize solute molecules; assume the membrane is semipermeable.

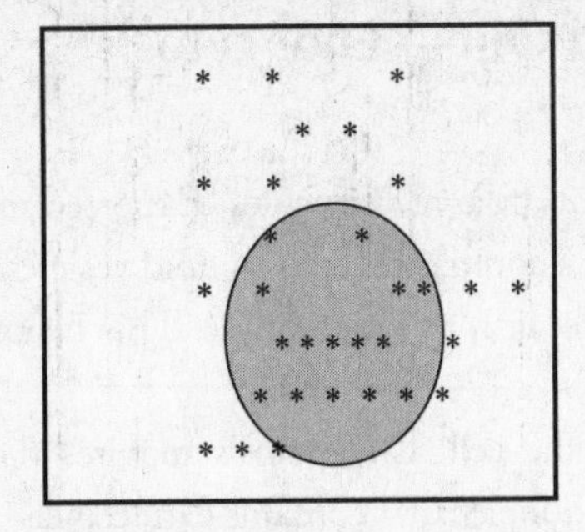

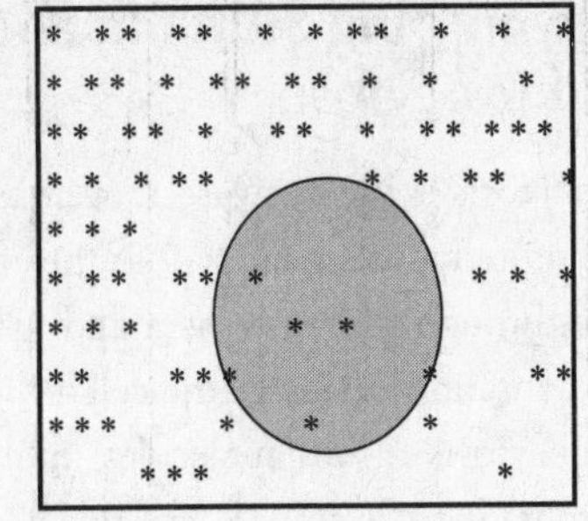

Passive and Active Transport

- Passive transport requires no input of energy; molecules simply move down their concentration gradient. Large molecules that cannot pass through the membrane are facilitated in diffusion by protein channels within the membrane.
- Active transport requires energy because molecules are being moved against their concentration gradient—these molecules are being concentrated either outside or inside of the cell.

THINK ABOUT IT

1. Fill out the following table to compare and contrast the following mechanisms of transport across membranes:

	Simple Diffusion	Facilitated Diffusion	Active Transport
Energy required?			
Direction (with respect to a concentration gradient)			
Membrane proteins			
Types of substances transported			

Moving "Big Stuff" In and Out— Endocytosis and Exocytosis

- Exocytosis is the outward movement of large molecules. Exported molecules, like proteins, are made within the cell and packaged into lipid vesicles that diffuse through the cytoplasm and fuse with the membrane. This fusion dumps the contents outside of the cell.
- Endocytosis moves molecules into the cell. Endocytosis requires the plasma membrane to invaginate, creating a "bubble" that contains extracellular material.
- Small endocytotic vesicles collect water and solutes; this process is called pinocytosis or "cell drinking." Larger vesicles form when extracellular receptors bind their ligands, then collect in one spot on the membrane and form a pit. This pit invaginates and internalizes both the receptor and ligand in a process called receptor-mediated endocytosis.
- Some cells can engulf very large items, like whole bacteria. The cellular membrane pushes out pseudopodia that surround and capture the item for the cell in a very large (1–2 mm) vesicle. This process is called phagocytosis, or "cell eating."

THINK ABOUT IT

1. What kinds of materials are transported by each of these processes?
   a. exocytosis
   b. phagocytosis
   c. receptor-mediated endocytosis

2. Draw a brief sketch of each of the above processes, with labels indicating key steps and structures.

Cool Fact: Exocytosis, Endocytosis, and Schizophrenia

As you'll learn in Chapter 12, the brain and the nervous system communicate cell-to-cell through the action of specialized molecules called neurotransmitters. Neurotransmitters are small molecules made in neurons and stored in vesicles in the tips of the axons. When a nerve signal arrives at the tip of an axon, it triggers the release of the neurotransmitter molecules through exocytosis into the synapse, the space between nerve cells. The neurotransmitter diffuses across the synapse and binds to its specific receptor so it can be endocytosed into the adjacent nerve cell and trigger another nerve signal. Nerve signal transmission, and thus brain function, depends on complementary processes of exocytosis and endocytosis.

Schizophrenia may be due to a defect in the transmission sequence triggered by the neurotransmitter dopamine. Some studies have indicated that compared to control patients, schizophrenics have altered levels of synthesis of a dopamine receptor in their cerebral cortex. Familial studies have suggested that inheriting a specific allele of one of the dopamine receptor genes may be associated with developing schizophrenia. An observation in patients with Parkinson's disease also suggests that defects in the dopamine signal may produce some of the delusional symptoms of schizophrenia. Parkinson's disease is linked to a lack of dopamine in the brain, but if Parkinson's patients take too much L-DOPA (a precursor molecule in the production of dopamine) then they show psychoses similar to those seen in schizophrenic patients, suggesting that too much dopamine in the synapse is a problem. Although we don't know the details of the dopamine transmission pathway—and specifically, how it is altered in schizophrenia—these few research findings hint at just how important exocytotosis and endocytosis can be in human health.

Chapter 5

TESTING, TESTING

Create a web that demonstrates the relationship between: active transport, integral proteins, peripheral proteins, facilitated diffusion, diffusion, and osmosis.

Vocabulary Review

Place each term from the following list into the appropriate column of the table below. Some terms may belong in more than one column.

active transport
cholesterol
communication
concentration
diffusion gradient
endocytosis
exocytosis
fluidity
fluid mosaic model
glycocalyx
hydrophilic
hydrophobic
intracellular
osmosis
passive diffusion
phagocytosis
phospholipids
plasma membranes
proteins
pumps
transport
transport of large molecules
vesicles

Membrane Components and Function	Gradients	Transport

PRACTICE TEST

Once you have finished studying this chapter, close your books, grab a pencil, and spend the next 15 to 20 minutes completing this practice test.

Compare and Contrast

For each of the following paired terms, write a sentence of comparison ("Both . . . ") and a sentence of contrast ("However, . . . ").

phospholipids/cholesterol

passive diffusion/facilitated diffusion

phagocytosis/pinocytosis

Matching

Match the following terms with their description. Each choice may be used once, more than once, or not at all.

a. integral protein
b. peripheral protein
c. cholesterol
d. phospholipid
e. carbohydrate ("sugar")

____Found as a bilayer
____Maintains membrane fluidity and serves as a patching compound
____Has ends that stick out of both sides of membrane
____Is both "water loving" and "water fearing"
____Serves as a binding site for proteins on the cell surface

Short Answer

1. Are cell membranes hydrophobic or hydrophilic? Why?

2. Why is it "easier" to move materials across membranes down a concentration gradient?

3. Why can't large materials move across the membrane by facilitated diffusion or active transport?

4. Explain why the movement of water across cell membranes should not occur.

Multiple Choice

Circle the letter that best answers the question.

1. Which of the following best describes the process of facilitated diffusion?

   a. movement of water from areas of high solute concentration to areas of lower solute concentration
   b. expenditure of energy to move solutes across a membrane against their concentration gradients
   c. expenditure of energy to move solutes across a membrane down their concentration gradients
   d. movement of solutes across a membrane down their concentration gradient with the involvement of membrane proteins
   e. movement of solutes across a membrane down their concentration gradient in the absence of membrane proteins

2. All of the following are true concerning the plasma membrane EXCEPT:

   a. It is composed of a phospholipid bilayer.
   b. It has integral and peripheral proteins.
   c. It forms the border between the cell and the external environment.
   d. Cholesterol helps add a rigid framework, or scaffold, to the membrane.
   e. All of these answers are true.

3. You have mutated an organism (Larklula) to produce chocolate candy within its cells. But Larklula is useless unless it can release the candy from its cells so that you can harvest it. Which of the following processes will Larklula need to use to accomplish this?

   a. osmosis
   b. phagocytosis
   c. exocytosis
   d. pinocytosis
   e. halitosis

4. Which of these is not a function of membrane proteins?

   a. structure
   b. recognition
   c. reproduction
   d. communication
   e. transport

5. Which of the following features explains why plant cells don't explode as a result of osmosis?

   a. Chloroplasts produce sugar, which prevents water from being drawn into the cell.
   b. Plant cells are all dead at maturity.
   c. Plant cells have a large reservoir, called a central vacuole, where they keep their water.

d. Plant cells have a rigid cell wall.
e. Actually, plant cells are more likely than animal cells to explode because of osmosis.

6. To build tissues it is necessary to bind cells together by attaching membrane proteins and phospholipids. Short chains of what kind of molecule form this glycocalyx?
   a. carbohydrate
   b. protein
   c. phospholipid
   d. nucleic acid

7. All of these are critical functions of cell membranes except:
   a. holding necessary substances within the cell
   b. preventing invasion of harmful material
   c. releasing energy for building new structures
   d. allowing passage of material into and out of the cell
   e. all of the above are critical functions

8. What is a concentration gradient?
   a. a difference in the density of substances from one side of a structure to the other
   b. a way of measuring depth perception
   c. a type of protein used to bring material into the cell
   d. the amount of energy required to run the sodium-potassium pump in active transport
   e. the amount of molecules within a cell

9. You are given two substances to study, one large and one small. Given what you have learned about the movement of material across membranes, which of the following methods would you predict could be used by cells to transport the larger substance?
   a. diffusion
   b. osmosis
   c. active transport
   d. exocytosis

10. You observe a cell as it takes up certain molecules. You notice that transport depends on the interaction of substance X with a membrane protein. You conclude that the mode of cell uptake being demonstrated is
    a. receptor-mediated endocytosis.
    b. facilitated diffusion.
    c. phagocytosis.
    d. pinocytosis.
    e. exocytosis.

11. Let's say you find yourself in an alternate universe, where lipid is the universal solvent and comprises approximately 70 percent of your body. What properties would the molecules that make up your cellular membranes have?

    a. They are likely to be lipid soluble.
    b. They are likely to be water soluble.
    c. They would serve as a semi-permeable membrane, allowing only certain materials to freely cross the membrane.
    d. Both a and c.
    e. Both b and c.

12. You decide to try your hand at canning pickles. You immerse freshly picked cucumbers in a solution that has a solute concentration twice that found in the cucumber cells. You allow your preparation to cure for several months in a sealed jar. When you open the jar later, you find that the fluid surrounding the "pickle" is more dilute than when you started. This change in concentration is due to

    a. water leaving the cucumber along its concentration gradient.
    b. water traveling against its concentration gradient.
    c. solute leaving the cucumber.
    d. solute entering the cucumber.

BUT, WHAT'S IT ALL ABOUT?

Here's a question to help you pull together what you've learned so far using this text. If you don't remember how to attack this type of question, check out the example in Chapter 1.

Back in Chapter 1, we considered how to define living things. We considered such features as assimilating and using energy, responding to the environment, reproducing themselves, and evolving from other living things. Whereas all of these definitions are helpful, I suggest that the simplest way to distinguish living and nonliving things is by the presence of plasma membranes (or one plasma membrane). Is this a reasonable statement and why?

1. **What kind of question is this?**
   This question asks you to defend the statement that only living organisms have plasma membranes.

2. **Collect the evidence.**
   What do we need to know to answer this question?
   We know that rocks, water, and soil lack plasma membranes. The object is to describe how the plasma membrane makes these functions (like responding to the environment) possible.

3. **Pull it all together.**
   Select several features of living organisms and provide examples of how the plasma membrane helps the organism accomplish these tasks on a cellular lever. You can use the following sentence to get you started: **Because plasma membranes allow living organisms to do the things they do, we can argue that the simplest definition of a living thing is something that has plasma membranes.**

6

CHAPTER 6

Life's Mainspring: An Introduction to Energy

Basic Concepts

- Living organisms capture energy from the Sun and use it to do complex activities. with some energy lost as heat.
- Energy is the capacity to move against an opposing force; energy can be transformed but is not lost or gained (first law of thermodynamics).
- Energy appears in biologic systems in the form of the molecule ATP; the energy stored in its chemical bonds can be transformed to cause movement, to transport molecules, or to make other chemical bonds.
- Enzymes are proteins that speed up and regulate energy transformations in living organisms.

Principles of Thermodynamics

- We define energy by its effect. Energy cannot be isolated, because it is an intrinsic property of a system. We distinguish the potential of a system to create change as potential energy, whereas energy seen as motion is kinetic energy.
- The ultimate source of energy for all living organisms is the Sun.
- The first law of thermodynamics allows energy to be transformed. Solar energy can be converted into chemical energy, and chemical energy can be converted into motion. However, energy cannot be created or destroyed.
- The second law of thermodynamics tells us that giving up energy means moving from an ordered state to a less-ordered state; entropy is a measure of the degree of order in a system.

THINK ABOUT IT

1. Do living things create energy, or transform it from one form to another?

2. Name a source of energy for living things.

3. Which form of energy (potential or kinetic) is represented by the following conditions?

___table sugar ready to ingest
___a diver about to dive off the 10m platform at the Olympics
___a gerbil running on its exercise wheel in its cage

4. If the diver in question (3b) ingests the table sugar, how will the energy in the sugar be transformed?

5. Which one are you more likely to see in action—kinetic or potential energy?

How Living Things Use Energy

- Energy is used by living systems to do three kinds of work: mechanical, transport, and synthetic.
- Each kind of work creates a local increase in the orderliness of the organism; such processes require the input of energy and are called endergonic reactions.
- Exergonic reactions release energy and create a decrease in the orderliness of the organism; this released energy powers the endergonic reactions as a coupled system.
- ATP (adenosine triphosphate) is a relatively unstable molecule that releases large amounts of energy when it loses a phosphate group, which can be harnessed to run an endergonic reaction. ATP doesn't store energy for long periods of time; instead, it transfers it from one molecule to another.

THINK ABOUT IT

1. You convert a pile of bricks into a greenhouse. The greenhouse is more organized than the pile of bricks. How can this be reconciled with the second law of thermodynamics?

2. List the three types of work observed in living things, and give an example of each.

   (a)

   (b)

   (c)

3. For the two reactions here, which is endergonic and which is exergonic?

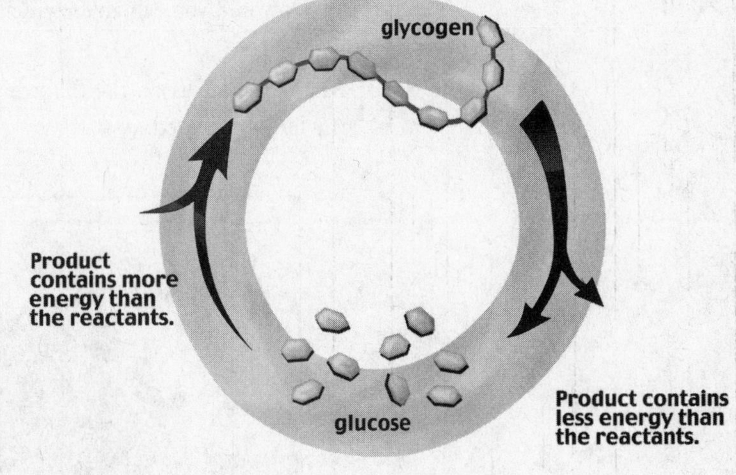

4. Based on their charges, which of the following hypothetical molecules would be predicted to be an unstable molecule that could, like ATP, be used to transfer energy?

   M+++ N+−+ O−+− P−−−

5. Use the following terms to fill in the diagram: endergonic, ADP, ATP, exergonic, P (phosphate):

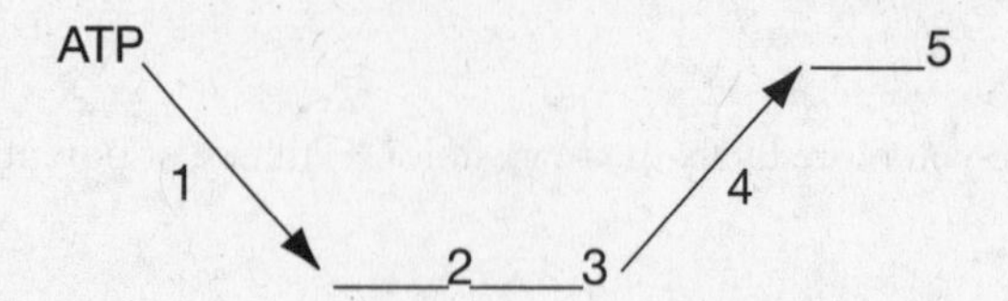

Efficient Energy Use in Living Things: Enzymes

- To carry out the complex activities of life, organisms must have a way to control the capture and release of energy.
- Specialized molecules called enzymes speed up virtually all biologic reactions by lowering the activation energy barrier.
- Enzymes are not consumed by biologic reactions; they are catalysts that are restored at the end of the reaction and can be reused.
- Enzyme activity can be controlled by the presence of regulator molecules to slow down or speed up reactions, in a process known as allosteric regulation.

THINK ABOUT IT

1. If a reaction is inherently "downhill," why does the cell require enzymes to be involved in chemical reactions?

2. Explain the difference between metabolism and a metabolic pathway.

3. a. Based on the name, predict the substrates of each of the following enzymes:

   (i) phospholipase
   (ii) DNAase
   (iii) cellulase

   b. What would you call an enzyme with RNA as its substrate?

4. a. Match the labels on the following energy diagram of a chemical reaction with the terms provided:

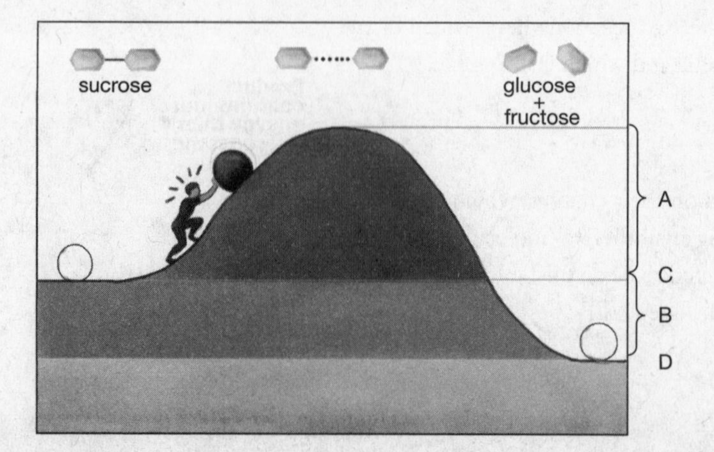

____ overall energy change of the reaction
____ energy of the products
____ energy of the reactants
____ activation energy

b. Which of these is altered by the activity of enzymes?

5. An enzyme has catalyzed a chemical reaction. Can it catalyze another reaction?

6. What types of macromolecules are enzymes?

7. You've always been told to take your vitamins. Based on what you know about enzymes, why do you think vitamins are important?

8. a. Which form of enzyme regulation involves a change of the enzyme's shape?

b. Why might a shape change affect enzyme activity?

9. Use Figure 6.11 to answer the questions that follow.

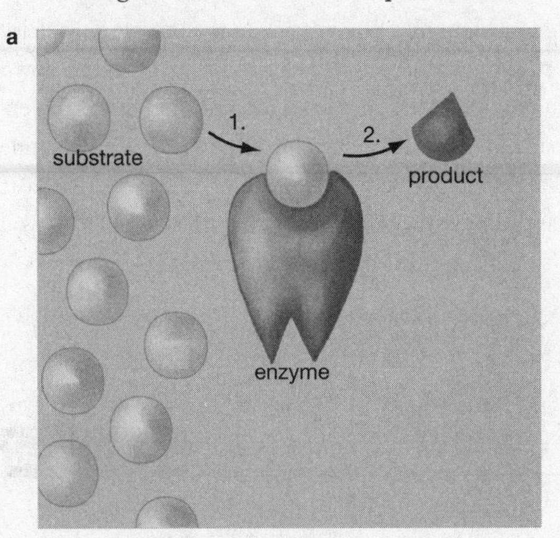

1. Substrate binds to enzyme.
2. Enzyme transforms substrate to product.

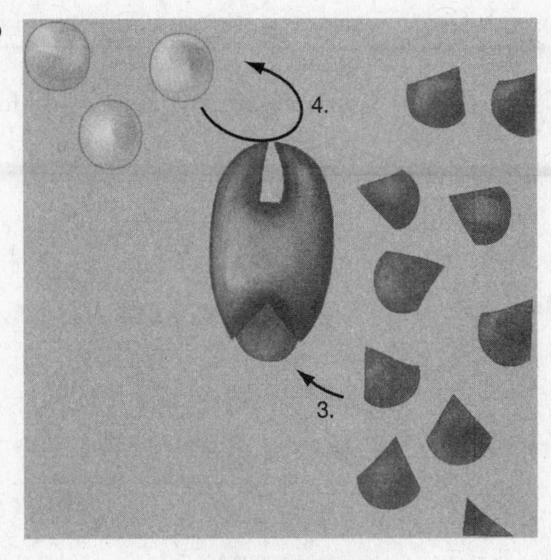

3. Product binds to a *different* site on the enzyme, causing the enzyme to change shape.
4. The new shape of the enzyme prevents it from binding to any more substrate.

a. Which site is the active site?

b. Which molecule is an allosteric inhibitor?

c. How does the inhibitor control enzyme action?

Cool Fact: Energy, Athletes, and Creatine

ATP is the energy currency of the cell. It moves energy from one reaction to another because it releases a lot of energy when one of its phosphate bonds is broken—and it's relatively easy to break the phosphate bond because it is not very stable, due to strong repulsive forces of the negatively charged oxygens in the phosphate groups. The flip side of this process is that you need to put in a lot of energy to create a phosphate bond in ATP. If ATP is mostly an energy transfer molecule, is there a more stable molecule that can store energy for a cell?

A specialized energy storage molecule exists in muscle, a tissue that experiences rapid demands for lots of energy. This molecule, creatine, can form a stable, high-energy phosphate bond by linking a phosphate group from ATP to a nitrogen (phosphocreatine). Muscle tissue has a reserve of phosphocreatine that can rapidly transfer this phosphate back to ADP and create lots of ATP when the muscle needs energy on short notice.

Athletes looking for an edge have turned to using creatine because—as the logic goes—if you can increase the amount of stored energy in the muscle, then you can extend the amount of time the muscle can work before it gets tired. Although it's true that by increasing the amount of creatine in the muscle (by ingesting or injecting creatine) you can increase the amount of phosphocreatine, this doesn't dramatically increase the amount of time a muscle can work before it fatigues. More phosphocreatine in the muscle does delay the switch from making ATP from phosphocreatine to making ATP by breaking down sugar molecules. This is useful for fueling activities that need a burst of energy (powerlifting, sprinting, or swinging a baseball bat), but the cell uses up all of its phosphocreatine within the first 30 seconds of activity. Endurance activities, meaning anything that lasts longer than 30 seconds, are not enhanced by extra phosphocreatine. Bottom line? If you want to be a better athlete, get more training, not more phosphocreatine.

Source: M. Kamber, M. Koster, R. Kreis, G. Walker, C. Boesch, and H. Hoppeler. 1999. Creatine supplementation—Part I: performance, clinical chemistry, and muscle volume. Med Sci Sport Exerc 31: 1763–1769.

Chapter 6

TESTING, TESTING

Create a web that demonstrates the relationship between potential energy, kinetic energy, sunlight, chemical energy, entropy, and transformation.

Vocabulary Review

Place each term from the following list into the appropriate column of the table below. Some terms may belong in more than one column.

ADP	endergonic	metabolism
ATP	entropy	protein
co-enzyme	exergonic	thermodynamics
coupled reactions	metabolic pathway	vitamin

Energy	Enzymes

PRACTICE TEST

Once you have finished studying this chapter, close your books, grab a pencil, and spend the next 15 to 20 minutes completing this practice test.

Compare and Contrast

For each of the following paired terms, write a sentence of comparison ("Both . . . ") and a sentence of contrast ("However, . . . ").

endergonic/exergonic

enzyme/co-enzyme

potential energy/kinetic energy

competitive inhibitor/allosteric inhibitor

Matching

Match the following terms with their description. Each choice may be used once, more than once, or not at all.

a. substrate
b. endergonic
c. activation energy
d. entropy
e. active site

___ Level of disorder in the universe
___ Where enzymes and substrates meet
___ What is changed by an enzyme-mediated reaction
___ Reaction requiring energy
___ Amount of energy required to begin a reaction

Short Answer

1. Describe the function of co-enzymes.

2. Why do we need to supply our bodies with new energy sources on a fairly regular basis?

3. Why is ATP essential to life?

4. Define kinetic energy.

Multiple Choice

Circle the letter that best answers the question.

1. In which of the following reactions would energy be created?

   a. enzyme
   b. entropy
   c. endergonic
   d. endless summer
   e. none of the above

2. Imagine that you attempt to pour sand from one cup to another. Which of the following terms best describes the sand that spills instead of landing in the second cup?

   a. exergonic
   b. endergonic
   c. entropy
   d. enchilada
   e. exercise

3. Which of the following terms would be used to describe the process by which plants convert the energy of sunlight into the energy of chemical bonds in food?

   a. entropy
   b. endergonic
   c. exergonic
   d. exhausting
   e. exhilarating

4. Which of the following terms is correct?

   a. adenosyl transphosphate
   b. andesine triphosphate
   c. adenosine tripolymer
   d. adenosine triphosphate
   e. none of the above

5. A coupled reaction

   a. involves a male enzyme as well as a female one.
   b. is the result of mixing acids and bases.
   c. is a set of exergonic and endergonic reactions.
   d. means that an enzyme is paired with its co-enzyme during the reaction.
   e. none of the above.

6. While serving as a visiting professor on the planet Freedonia, you discover a life-form that does not use ATP as its primary energy transfer molecule. Instead, the organism uses michelerene phosphate (MiP). Michelerene phosphate stores energy by adding positively charged michelerene groups (Mi+) to the backbone phosphate. Which of the following would have the highest energy state (the greatest potential energy)?

   a. MiP+
   b. MiP++
   c. MiP+++
   d. MiP2^{-}
   e. MiP^{-}

7. The process of converting MiP+++ to MiP++ (see question 6) is

   a. exergonic
   b. endergonic
   c. potenitialic
   d. investive
   e. endoscopic

8. How does the cell use ATP to drive uphill reactions?

   a. by using energy to both make ATP and drive the uphill reaction
   b. by converting ATP to ADP, which releases energy to drive the uphill reaction
   c. by converting ADP to ATP, which releases energy to drive the uphill reaction
   d. by using enzymes that make the uphill reaction become downhill
   e. a and c

9. Why is ATP such a good energy currency molecule?

   a. It is very stable.
   b. Its negatively charged phosphate groups repel one another.
   c. It has three positive charges that repel one another, forming an unstable molecule.
   d. It has double covalent bonds, which store twice as much energy.
   e. All of the above are correct.

10. What part of a reaction is changed by an enzyme?

    a. the energy of the reactants
    b. the activation energy

c. the energy of the products
d. the energy difference between the products and the reactants
e. the alteration energy

11. At which site does an allosteric inhibitor bind?

    a. the active site of an enzyme
    b. the active site on the products
    c. the active site on the reactants
    d. a site on the enzyme away from the active site
    e. a site on the reactants not affected by the chemical reaction

12. The enzyme lactase participates in what reaction?

    a. the breakdown of lactose
    b. the formation of fructose
    c. the breakdown of glucose
    d. the formation of protein

13. A number of diseases like rickets, beriberi, and scurvy have been linked to vitamin deficiencies. How could these deficiencies cause illness?

    a. by decreasing cellular ADP
    b. by decreasing the efficiency of cellular enzymes
    c. by increasing the reproduction of bacteria that produce the vitamins
    d. by causing the cellular nucleus to dissolve

BUT, WHAT'S IT ALL ABOUT?

Here's a question to help you pull together what you've learned so far using this text. If you don't remember how to attack this type of question, check out the example in Chapter 1.

This chapter begins with a discussion of the first and second laws of thermodynamics. Re-read the section about these laws and consider the following observations:

1. Energy cannot be created or destroyed, so how can organisms make the energy they need to move, reproduce, and interact with the environment?
2. Organisms are made of cells, which are highly ordered structures. Is this a violation of the second law of thermodynamics?

1. **What kind of question is this?**
   This is a variation on the "defend a position" question. You are being asked to explain why two observations from the natural world that appear to be in violation of the laws of thermodynamics are truly not.

2. **Collect the evidence.**

   a. For your benefit, write down the two laws of thermodynamics in question and consider what it means to "transform energy" and "tend to disorder."
   b. Consider what it means to "make" energy in a cell – is it really creating energy or transforming it?
   c. Likewise, when we think of disorder in a "system," what defines a system?

3. **Pull it all together.**

   Plan to answer the questions in order, so you will need two paragraphs. Start with summarizing the laws of thermodynamics and then provide examples of how living organisms do not violate these laws. You can use the following topic sentences to get you started.

   According to the first law of thermodynamics, energy cannot be created or destroyed only transformed.

   The second law of thermodynamics states that the universe tends to become more disordered during these energy transformations; a state known as an increase in entropy.

7

CHAPTER 7

Vital Harvest: Deriving Energy from Food

Basic Concepts

- ATP is generated by the fall of energized electrons through various reduction-oxidation reactions.
- Three metabolic pathways extract energy from food in complex organisms—glycolysis, the Krebs cycle, and the electron transport chain.
- Breaking the bonds of ATP supplies energy for metabolic reactions.

Oxidation-Reduction Reactions

- Oxidation-reduction (redox) reactions always happen in pairs; molecules that attract electrons strongly accept electrons from high-energy molecules. The molecule that accepts the electrons becomes reduced while the high-energy donor becomes oxidized.
- Electron transfer proceeds down the energy hill; that is, we start with a high-energy molecule donating electrons to a lower-energy acceptor molecule. The molecule donating electrons becomes oxidized, whereas the acceptor becomes reduced. Oxygen is not required for redox reactions; any molecule that donates electrons can be oxidized, and any molecule that can accept electrons can be reduced.

THINK ABOUT IT

1. Is the reaction converting ADP to ATP exergonic or endergonic?

2. What happens in a redox reaction?

3. Label each of the following reactions as an oxidation (O) or a reduction (R):

   ___$NADH \rightarrow H+ + e^- + NAD+$

   ___$2H+ + 2e^- + O \rightarrow H_2O$

   ___$6CO_2 + 6H_2O \rightarrow C_6H_{12}O_6$

The Three Stages of Cellular Respiration: Glycolysis, The Krebs Cycle, and the Electron Transport Chain

- Glycolysis is the obligatory first stage in energy harvesting; it exists in all organisms. Though not very efficient, producing only 2 ATP for each glucose oxidized, it provides essential substrates for the Krebs cycle and the electron transport chain—2 NADH and 2 pyruvic acid molecules per glucose molecule oxidized.
- The Krebs cycle and the electron transport chain (ETC) are evolutionarily more recent additions to the energy-harvesting machinery. These two stages require oxygen and produce an enormous amount of energy (32 ATP) from a single glucose molecule. These reactions occur within the mitochondria, the powerhouse of the cell.
- The Krebs cycle, also called the citric acid cycle, accepts the carbons from pyruvic acid as the high-energy intermediate called acetyl CoA, and oxidizes them completely to carbon dioxide (CO_2), which we exhale. The Krebs cycle also "saves" the energy of these oxidation reactions in the form of the reduced electron carrier molecule, NADH.
- The ETC allows the high-energy NADH molecule to donate its electrons to other electron carriers that use the energy released in this downhill transfer of electrons to pump protons ($H+$) up an energy hill against their concentration gradient. Releasing these protons to flow down their concentration gradient releases the energy needed to make ATP.
- Molecules other than glucose can be oxidized to produce energy, but all of these alternative oxidation pathways eventually feed into glycolysis, the Krebs cycle, or the ETC.

THINK ABOUT IT

1. Draw a diagram of the cell showing where the three stages of energy harvesting take place.

2. What valuable molecules are produced during glycolysis?

3. Create a flowchart or diagram to show the key steps of glycolysis. Note on your diagram your starting molecule's structure and the structure of the final molecule. Indicate where in the process ATP must be invested, and where ATP is produced. Indicate where NADH is produced. Figure out which intermediates represent high-energy molecules—those that can donate either a phosphate group or electrons to an acceptor molecule.

4. During which stage of cellular respiration is oxygen consumed?

5. Which stage of cellular respiration yields the most ATP?

6. In which compartment of the cell do the reactions of the Krebs cycle occur?

7. Which reduced electron carriers are produced during the Krebs cycle?

8. Fill out the following table describing the inputs and outputs of the Krebs cycle.

Input Molecules	Output Molecules

9. What happens to the electrons carried by NADH and $FADH_2$ when they arrive at the ETC?

10. The ETC transports electrons and pumps protons. How are these events related to ATP synthesis?

11. What reduced form of oxygen is produced during the final stage of respiration? (Hint: Think about what happens to oxygen at the end of the ETC.)

12. If fats (triglycerides) are being used as an energy source, where do they enter the cellular respiration pathway?

13. The storage carbohydrate formed by the assemblage of glucose molecules is called...

Cool Fact: Red Blood Cells and Glycolysis

Red blood cells have a single function—to carry oxygen to all the tissues in the body, so that cellular respiration can occur. All tissues, that is, except one—themselves. Red blood cells lack mitochondria and depend completely on glycolysis to supply their energy needs. Red blood cells also depend on glycolysis because one of the intermediates in the pathway can be used to make a molecule that helps hemoglobin release oxygen into the tissues more efficiently. One pathway serves two purposes, as an interesting example of how efficient evolution can sometimes be.

Chapter 7

TESTING, TESTING

Create a web that defines the relationships between mitochondria, cytoplasm, redox, ATP, ETC, glycolysis, Krebs cycle, and electrons.

Vocabulary Review

Place each word in the appropriate column of the table. Keep in mind that some words may belong in more than one column.

acetyl Co A	citric acid cycle	H+ pumps	NADH
ADP	FAD+	H_2O	oxygen
ATP	$FADH_2$	inner membrane	pyruvic acid
ATP synthase	fructose-6-phosphate	mitochondria	"sugar splitting"

Glycolysis	Krebs Cycle	Electron Transport Chain

PRACTICE TEST

Once you have finished studying this chapter, close your books, grab a pencil, and spend the next 15 to 20 minutes completing this practice test.

Compare and Contrast

For each of the following paired terms, write a sentence of comparison ("Both . . . ") and a sentence of contrast ("However, . . . ").

NADH/$FADH_2$

oxidation/reduction

mitochondria/cytoplasm

pyruvic acid/citric acid

Matching

Match the following terms with their description. Each choice may be used once, more than once, or not at all.

a. glycolysis
b. electron transport chain
c. acetyl CoA formation
d. Krebs cycle
e. oxidation
f. reduction

___ Process by which an electron is lost

___ A 3-carbon molecule is converted to a 2-carbon molecule; a molecule of CO_2 is released

___ Process by which an electron is gained
___ Stage at which oxygen is required
___ Takes place in inner mitochondrial compartment
___ Shared by all living organisms

Short Answer

1. What is the evolutionary significance of the Krebs cycle?

2. Explain the relationship between the terms redox and coupled reactions.

3. Explain how oxygen is used during glycolysis.

4. Why do we exhale CO_2?

Multiple Choice

Circle the letter that best answers the question.

1. Why are H+ ions pumped across the inner mitochondrial membrane?

   a. So they can get to the other side.
   b. They are the waste products of cellular respiration.
   c. Their movement back across the membrane generates ATP.
   d. It is the step that makes the Krebs cycle work as a cycle.
   e. It generates acetyl CoA.

2. Why is the Krebs cycle a cycle?

   a. because the most important molecule, oxaloacetate, is circular
   b. because while the rest of cellular respiration happens during the day, it takes place at night
   c. because the first molecule in the pathway is also the last
   d. because it takes place in the mitochondria, which are round
   e. because that is the name that Krebs chose

3. Which stage yields the greatest amount of ATP?

   a. Krebs cycle
   b. acetyl CoA formation
   c. electron transport chain
   d. glycolysis
   e. oxidation

4. Which stage is evolutionarily the oldest?

   a. Krebs cycle
   b. acetyl CoA formation

c. electron transport chain
d. glycolysis
e. oxidation

5. Which stage takes place in the cytoplasm?

a. Krebs cycle
b. ATP synthase
c. electron transport chain
d. glycolysis
e. oxidation

6. Which stage releases the first molecule of CO_2?

a. Krebs cycle
b. acetyl CoA formation
c. electron transport chain
d. glycolysis
e. oxidation

7. Which of the following best describes an oxidation reaction?

a. A substance gains an oxygen atom.
b. A substance gains electrons.
c. A substance gives an oxygen atom to another substance.
d. A substance donates electrons to another substance.
e. none of the above

8. What is the correct order of the stages of cellular respiration?

a. Krebs cycle, electron transport chain, glycolysis
b. electron transport chain, Krebs cycle, glycolysis
c. glycolysis, Krebs cycle, electron transport chain
d. glycolysis, electron transport chain, Krebs cycle
e. Krebs cycle, glycolysis, electron transport chain

9. Which molecules can be oxidized to generate ATP?

a. glucose
b. fats
c. proteins
d. all of the above
e. none of the above

10. Back on the planet Freedonia, with infinite resources available to you, you decide to try implanting mitochondria into bacterial cells. Assuming your experiment is successful, what results would you predict?

    a. no change
    b. an increase in ATP yield/glucose molecule
    c. an increase in oxygen usage by the bacteria
    d. a decrease in oxygen usage by the bacteria
    e. both b and c

11. Suppose you were to then supply the cells with the means of making acetyl CoA, what would you expect to see?

    a. no change
    b. an increase in ATP yield/glucose molecule
    c. an increase in oxygen usage by the bacteria
    d. a decrease in oxygen usage by the bacteria
    e. both b and c

12. You and your brother are racing each other home. The first one home will get the last piece of your mom's delectable chocolate cake. As you run, you feel a burning sensation in your leg muscles, which you know is caused by lactic acid fermentation. Why do your muscle cells switch from aerobic respiration to anaerobic respiration which generates fewer ATP per glucose molecule used?

    a. Aerobic respiration is expensive and periodically the cells "take a break" and use anaerobic respiration for a period of time.
    b. Muscle cells don't have mitochondria and therefore are incapable of performing aerobic respiration.
    c. Aerobic respiration takes longer to get started and so muscle cells switch to fermentation for short, intense activities.
    d. Anaerobic respiration can be run using muscle proteins (actin and myosin) as an energy source so the muscles can perform longer.

BUT, WHAT'S IT ALL ABOUT?

Here's a question to help you pull together what you've learned so far using this text. If you don't remember how to attack this type of question, check out the example in Chapter 1.

Every November, millions of Americans sit down to a feast, probably the largest, single meal they will eat all year. And after this gorge, most of us leave the table to watch televised football or fall asleep. Based upon your new knowledge of cellular respiration, and your old knowledge of the structures of biomolecules, describe how we use the turkey, dressing, vegetables and pumpkin pie we have just eaten.

1. **Type of Question.**
   You are being asked to apply your knowledge to solve a "problem", that is, figure out an explanation for a novel scenario.

2. **Collect the Evidence.**
   You already know from past experience the major food groups correspond to our major classes of biomolecules – proteins, carbohydrates and fats. Using what you have learned about deriving energy from food, describe the fate of the molecules consumed during Thanksgiving dinner.

3. **Pull it all Together.**
   The question has given you two key points while describing this scenario: the quantity of food consumed (large) and the level of activity following this meal (low). You must consider both of these points in framing your answer. You may use the following topic sentence to get started:

   Our Thanksgiving feast is comprised of various biomolecules: protein and fat in the turkey, mostly carbohydrates in the dressing and vegetables, and mostly fats and simple sugars in the pie.

CHAPTER 8

The Green World's Gift: Photosynthesis

Basic Concepts

- Photosynthesis captures the light energy of the Sun and stores it in the carbohydrates that plants produce. Because most organisms depend on plants as a source of fuel (directly or indirectly), photosynthesis makes life on Earth possible.
- Photosynthesis has two stages—the first saves light energy as energetic electrons in NADPH and releases oxygen, and the second delivers the energy in the electrons of NADPH to a 3-carbon sugar and CO_2 to make food.
- Alternate metabolic cycles make photosynthesis possible under difficult environmental conditions.

Photosynthesis and Energy

- Photosynthesis creates the food for all organisms from sunlight, water, and CO_2—food that is broken down in the energy-harvesting reactions of glycolysis, the Krebs cycle, and the electron transport chain (cellular respiration).
- Photosynthesis generates O_2 as a by-product.
- Photosynthesis and cellular respiration are opposing reactions; photosynthesis is an endergonic process that stores solar energy as chemical energy in food, whereas cellular respiration is an exergonic process that releases chemical energy for mechanical or chemical work.

THINK ABOUT IT

1. Could eukaryotic cells have evolved without photosynthesis? Why or why not?

2. Fill in the blanks in this diagram to show the relationship between photosynthesis and cellular respiration.

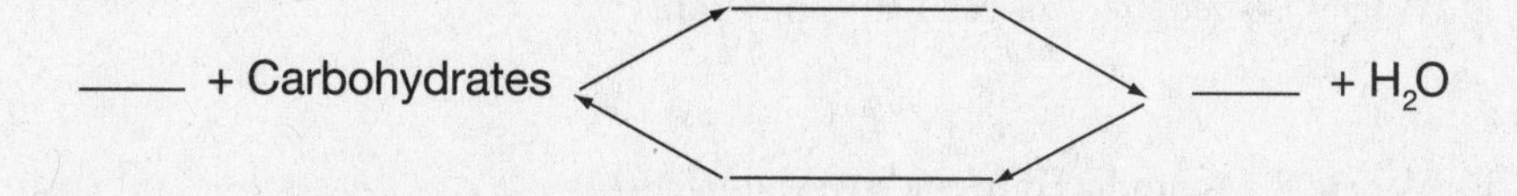

3. Which gases are used and produced during photosynthesis?

4. Photosynthesis relies on electrons. Where do the electrons come from?

Two Stages of Photosynthesis

- The photo stage of photosynthesis occurs within the membrane networks of a specialized cellular organelle, the chloroplast. Chloroplasts contain pigments that allow the capture of solar energy and the conversion of that energy into chemical energy through a series of redox reactions.
- The synthesis stage of photosynthesis uses the energized electrons and the energy produced in the photo stage to reduce CO_2 and make carbohydrates.
- The working units of photosynthesis are two photosystems. These photosystems transfer electrons taken from water and energized by absorbing solar energy through a series of redox reactions, converting that solar energy into reducing power stored as NADPH and into a proton gradient that can make ATP.
- The working unit of carbohydrate synthesis is the enzyme rubisco. Rubisco fixes carbon from atmospheric CO_2 as 3-phosphoglyceric acid, a 3-carbon sugar that can be used to make glucose.

THINK ABOUT IT

1. How do the gases used and produced in photosynthesis enter and leave the plant?

2. a. Which plant organelles convert the energy in sunlight into carbohydrates?

   b. Draw a plant cell and label the following: chlorophyll, accessory pigments, chloroplast, thylakoid, grana, and stroma.

3. What is the function of the light-dependent reactions?

4. What provides the energy to boost an electron up the energy hill from photosystem II?

5. Where do the electrons from photosystem II end up?

6. Eukaryotic cells need O_2 for cellular respiration.

   a. Why is O_2 required for cellular respiration?

   b. Where is O_2 produced in the environment?

7. What are two important products of the light-dependent reactions? Where are they produced?

8. Compare the electron transfer process of photosynthesis with the electron transfer chain of mitochondria. Make a list of the features they have in common and those that differ.

Differences	
Similarities	

9. a. What is carbon fixation?

   b. What is the photosynthetic pathway of carbon fixation called?

10. What is rubisco?

11. Balance the following equation:

    $3\ CO_2 + 3$ _____ $\rightarrow 6\ [3PGA] \rightarrow \rightarrow 6$ _____

12. Build a flowchart that follows the pathway of a CO_2 molecule through photosynthesis and cellular respiration. Repeat the process for an O_2 molecule.

Alternative Pathways of Photosynthesis

- In photosynthesis, there is a "glitch" known as photorespiration. Photorespiration occurs because rubisco can bind O_2 as well as CO_2; however, when it binds O_2, energy is used but no sugars are made.
- Alternative pathways have developed in plants in warm, wet and warm, and dry climates to get around this wasteful process.
- C_4 plants fix CO_2 using a different reaction, then release CO_2 into the bundle-sheath cells for use in the Calvin cycle, at the cost of ATP energy.
- CAM plants—common in hot, dry climates—fix CO_2 using the C_4 pathway at night and wait until light is available before proceeding with the Calvin cycle.

THINK ABOUT IT

1. Is photorespiration an efficient source of carbohydrates for plants? Why or why not?

2. If global levels of CO_2 are increasing, would you predict photorespiration to increase or decrease?

3. a. What types of plants use the C_4 pathway of carbon fixation?

   b. Why is this pathway found in these plants?

4. You find an arctic tundra plant that appears to use the C_4 pathway. Why is this a surprising finding?

5. Describe the features of the C_4 pathway of photosynthesis that distinguish it from the usual pathway of photosynthesis.

6. Why do tropical plants often use variants of the photosynthetic pathway?

7. How does the CAM pathway differ from the C_4 pathway?

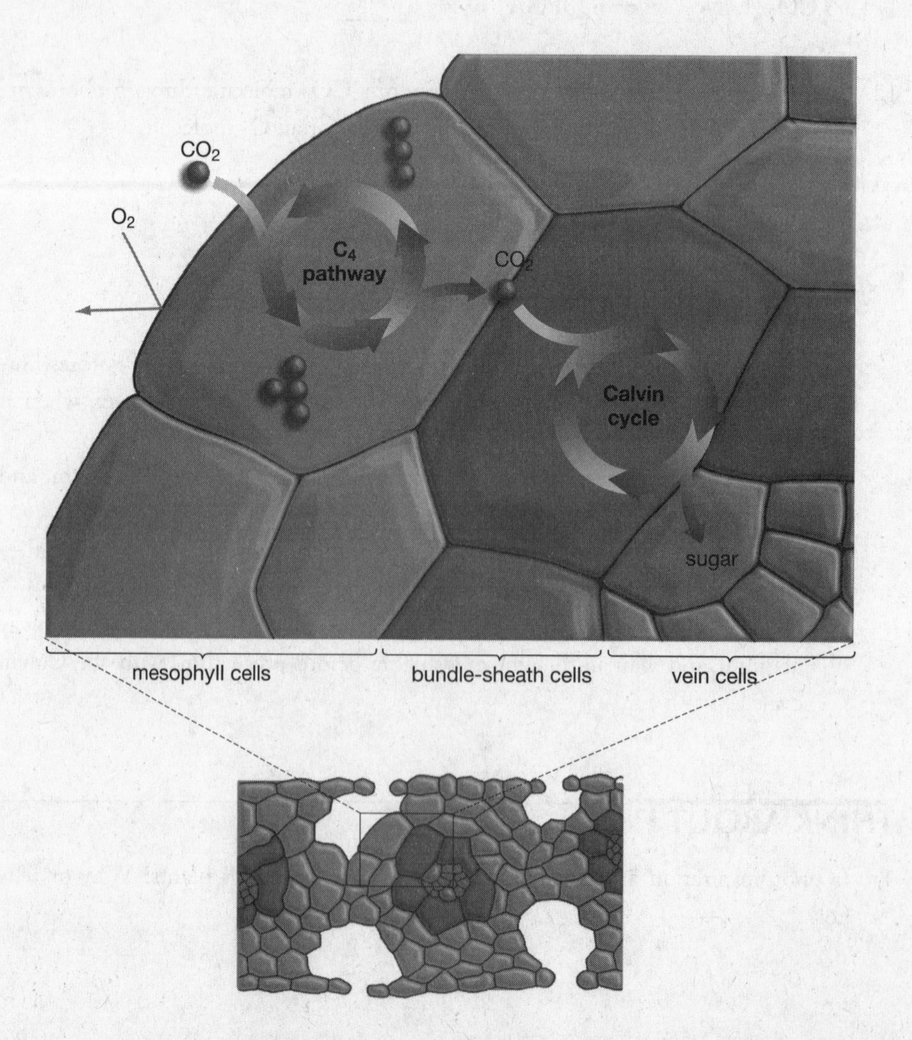

Cool Fact: The Gaia Hypothesis

Did chemical activity on an ancient Earth produce an environment that allowed living organisms to appear, and then those organisms adapted to the conditions? Or, did the physical environment change in response to the appearance of living organisms, so that living organisms and their physical environment "co-evolved"? The Gaia hypothesis (Gaia means "Earth goddess") was proposed by James Lovelock, a physical scientist, and Lynn Margulis, a microbiologist, to explain the stability of the Earth's atmosphere. Our atmosphere—with its high oxygen, low carbon dioxide, and moderate temperatures—has been stable for over 3.6 billion years, indicating that some process exists to keep it in balance. The Gaia hypothesis suggests that cycles of nutrient flow (carbon, nitrogen, oxygen), moving from the atmosphere to organisms to soil and back to the atmosphere, are interconnected and work to regulate each other to maintain a biosphere optimized for life on Earth. We do not yet know how these different cycles are coordinated and regulated, but microbial organisms are good candidates for the "sensor" that keeps the cycles in balance.

If all of these cycles are coordinately regulated, then big changes in any one nutrient could make the entire system less stable. This is part of the concern about the increasing levels of CO_2 in the atmosphere generated by human activity and the loss of rain forests. The photosynthetic activity of Earth's forests fixes 1015 grams of carbon a year using atmospheric CO_2. But as we continue to clear rain forests and generate CO_2 by burning fossil fuels, we may disrupt all of the geochemical cycles that keep our environment stable, causing large-scale climatic changes (global warming, increased storm activity, drought, and so forth). Evidence is mounting that human activity is primarily responsible for a warming of Earth's climate. The question remains, just how much can physical conditions change and still support living organisms?

Chapter 8

TESTING, TESTING

Create a web that demonstrates the relationship between ATP, NADPH, light-dependent reactions, light-independent reactions, CO_2, H_2O, Calvin cycle, CAM

Vocabulary Review

Place each term from the following list into the appropriate column of the table. Some terms may belong in more than one column.

accessory pigments	CO_2	light	photosystem II
C_4	G_3P	NADPH	rubisco
CAM	glucose	O_2	RuBP
chloroplast	H_2O	photosystem I	stroma

Light-Dependent Reactions	Light-Independent Reactions

PRACTICE TEST

Once you have finished studying this chapter, close your books, grab a pencil, and spend the next 15 to 20 minutes completing this practice test.

Compare and Contrast

For each of the following paired terms, write a sentence of comparison ("Both . . .") and a sentence of contrast ("However, . . .").

CAM/C_4 photosynthesis

chloroplast/chlorophyll

photorespiration/photosynthesis

stroma/grana

Matching

Match the following terms with their description. Each choice may be used once, more than once, or not at all.

a. photosynthesis
b. cell respiration
c. light-dependent reactions
d. light-independent reactions
e. photorespiration
f. ATP

___ Releases the energy stored in the chemical bonds of food
___ The "synthesis" reactions of photosynthesis
___ When rubisco binds with oxygen instead of carbon dioxide
___ Uses the energy of the Sun to build carbohydrates
___ Its products are used in the Calvin cycle
___ The currency of cellular energy

Short Answer

1. Explain the significance of RuBP.

2. What stage of photosynthesis uses chlorophyll?

3. What is the function of NADPH?

4. Explain the function of each part of the first half of the photosynthesis equation. That is, how are CO_2, H_2O, and energy used to build sugar?

Multiple Choice

Circle the letter that best answers the question.

1. What is photorespiration?

   a. when rubisco binds O_2, not CO_2, and no growth occurs
   b. cellular respiration in plant cells
   c. synthesis of carbohydrates from CO_2
   d. cellular respiration in chloroplasts
   e. none of the above

2. Why is the C_4 pathway called C_4?

   a. because it involves four steps
   b. because it initially fixes CO_2 into a 4-carbon molecule
   c. because only four types of plants have this pathway
   d. because it involves the four C's (cut, clarity, . . .)
   e. because four molecules of CO_2 are fixed into carbohydrate

3. What is the initial electron donor in photosynthesis?

   a. O_2
   b. rubisco
   c. NADPH
   d. H_2O
   e. PSI

4. What are the products of photosynthesis?

   a. plant material, H_2O, and NADP+
   b. CO_2 and plant material
   c. CO_2 and O_2
   d. O_2 and plant material
   e. minerals, CO_2, and H_2O

5. Which of the following is (are) true about the relationship between the Calvin and Krebs cycles?

   a. One uses CO_2 while the other produces it.
   b. Both are named for the same man.
   c. Both are involved in energy transformation.
   d. Both produce ATP.
   e. Both a and c are correct.

6. All of the following terms refer to the light-dependent reactions except

   a. photosystem I
   b. photosystem II
   c. NADPH
   d. RuBP
   e. chlorophyll

7. Which of the following is equal to one-half of a glucose molecule?

   a. ATP
   b. NADPH
   c. G3P
   d. STP
   e. ABC

8. Stroma is

   a. the collection of molecules that light strikes first.
   b. the organelle where photosynthesis takes place.
   c. the opening in leaves that allows the passage of water and carbon dioxide.
   d. the liquid found inside the chloroplast.
   e. an Italian pasta dish.

9. In order to begin the Calvin cycle, RuBP must combine with ____________ and rubisco.

   a. water
   b. ATP
   c. sunlight
   d. carbon dioxide
   e. ruminants

10. Which of the following is the process that ultimately feeds the world?

    a. decomposition
    b. photosynthesis
    c. respiration
    d. reproduction
    e. regeneration

11. As you finish up your sabbatical on the planet Freedonia, you hear of a type of photosynthetic plants that cannot make rubisco and therefore cannot complete the Calvin cycle. You correctly think to yourself,

    a. "They must not have chloroplasts."
    b. "They can't really exist without the ATP produced in the Calvin cycle."
    c. "They are probably less complex, unicellular organisms like bacteria, due to their lack of energy-storage capability."
    d. "They are likely to be very large, multicellular organisms like Sequoia trees because of their rapid ATP turnover."
    e. "Plants on Earth don't need rubisco either."

12. Which of the following might be used as proof that global warming is occurring?

    a. a decrease in photorespiration rates
    b. an increase in the number of species with C4 adaptations
    c. a darkening in color of photosynthetic pigments
    d. a decrease in the cooking time required for fresh picked vegetables

13. Which of the following is NOT found in both plant and animal cells?

    a. ATP
    b. G3P
    c. Mitochondria
    d. Glucose
    e. All of these are found in both types of cells.

BUT, WHAT'S IT ALL ABOUT?

Here's a question to help you pull together what you've learned so far using this text. If you don't remember how to attack this type of question, check out the example in Chapter 1.

In this chapter we have explore how plants make carbohydrates (their food and ours) from nothing more than water, CO_2, and the energy of sunlight. In the previous chapter we saw how we (and plants) can extract this energy in the process of cellular respiration. Despite the differences in the function of these two pathways, they share a remarkable number of features. Describe how photosynthesis differs from cellular respiration, and how the two pathways are the same.

1. **Type of Question.**
   This is a compare and contrast question.

2. **Collect the Evidence.**
   Make a two-column table. List the similarities in one column, and the differences in the other. Use the information in your textbook to collect your evidence.

3. **Pull it all Together.**
   Because the question phrases the problem as "despite the differences . . . share . . . features," we would choose to point out the differences first and then provide examples of the similarities. On the surface, photosynthesis and cellular respiration are opposing pathways. You may use a starting sentence such as, **"On the surface, photosynthesis and cellular respiration are opposing pathways"** to introduce the differences and follow up with a sentence beginning with **"Despite these differences . . . "**, or **"However, both pathways share . . . "**

9

CHAPTER 9

Introduction to Genetics: Mitosis and Cytokinesis

Basic Concepts

- DNA encodes all of the information necessary for an organism to develop, grow, repair itself, and duplicate itself.
- Information is stored within the sequence of nitrogen-containing bases, a gene, and each gene specifies a different protein responsible for executing a cellular function.
- Duplication and transfer of genetic information occur during the cell cycle, a multi-stage process that involves: copying all of the genes (interphase) and equal division of chromosomes (mitosis), followed by division of the cell into two daughters (cytokinesis).

DNA Structure and Function

- Just as words store information using a sequence of letters, DNA stores information chemically as sequences of the nitrogen-containing bases adenine, thymine, cytosine, and guanine.
- DNA's double helical structure provides a mechanism for self-duplication and for information transfer; each strand of the helix can serve as a template, either for DNA replication or translation into a messenger RNA (mRNA).
- Genes are DNA sequences that specify the structure of a protein. Many genes are specified along the length of the DNA and protein complexes known as chromosomes. All of an organism's genes, packaged within multiple chromosomes, constitute its genome.

THINK ABOUT IT

1. a. Which of the following best represents the relationship between genes and the genome?

   a. pancakes and breakfast

   b. cars and streets

   c. books and libraries

   b. Explain why you chose the answer you did.

2. a. The instructions for what type of macromolecule are contained in DNA?

   b. How does DNA transmit instructions?

3. What does every individual inherit from his or her parents?

4. Place the following key macromolecules in the path of protein synthesis in their proper locations: DNA, ribosome, mRNA, amino acid, protein.

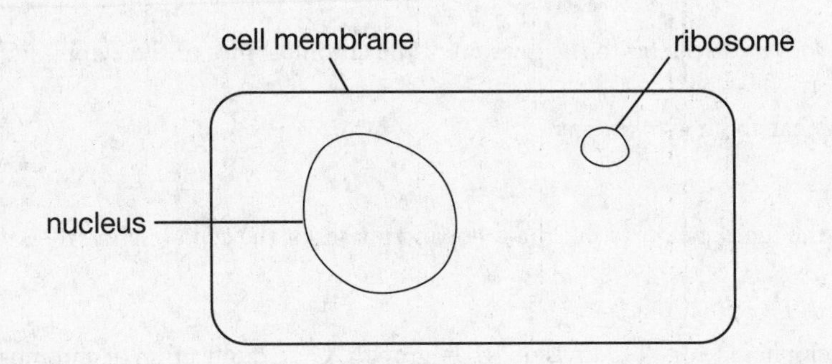

5. Why do chromosomes come in matched pairs?

6. Draw a chromosome before and after DNA replication has taken place.

7. a. How many homologous pairs of chromosomes are found in each human cell?

   b. How many pairs of autosomes are found in each human cell?

   c. How many pairs of human sex chromosomes are found in human cells?

   d. How do the sex chromosomes compare between human males and females?

Mitosis and Cytokinesis

- Mitosis and cytokinesis occur as part of the sequence of events that define cell function—the cell cycle. The cell cycle consists of two parts—a long interphase, during which the cell grows and carries out its specific functions (including replicating its DNA), and a shorter mitotic phase, during which the duplicated DNA is evenly divided between two daughter cells created as the parental cell splits.
- The mitotic phase is subdivided into four parts—prophase, metaphase, anaphase, and telophase.
- Prophase and metaphase prepare the cell to divide its genetic material equally. Specialized microtubule structures form during prophase that define two cellular poles and help align the duplicated chromosomes (each composed of sister chromatids) along an equatorial plane during metaphase.
- Chromosomes are pulled apart during anaphase and moved toward the opposite poles along the microtubules.

- Cytokinesis—cell division—occurs during telophase. Cytokinesis occurs as the cell membrane invaginates around the equatorial plane, pinching the original cell in two. Plant cells must also assemble a cell wall along this division furrow as the plasma membrane pinches inward.

THINK ABOUT IT

1. What is the primary function of cell division?

2. a. Does mitosis refer to the division of the chromosomes or the cytoplasm?

   b. What about cytokinesis?

3. Use the figure below to put the following phases of the cell cycle in order.

   S, G_1, M, G_2

Label each phase (A = cell division, B = growth, C = duplication of chromosomes).

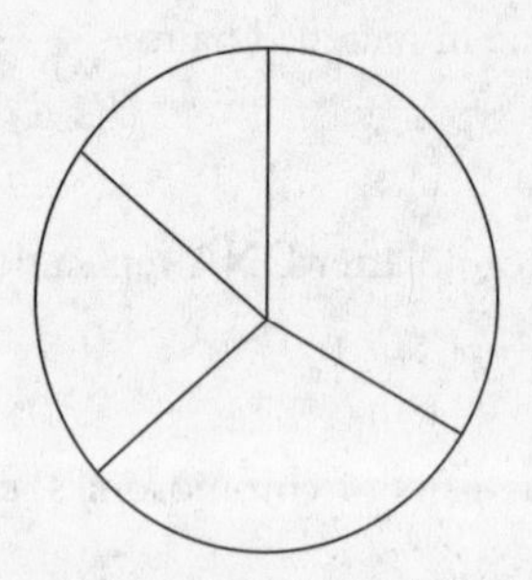

4. Fill out the following table describing the key events both within the nucleus and in the cytoplasm of each stage of mitosis and cytokinesis.

Stage	Key Events
Prophase	
Metaphase	
Anaphase	
Telophase	
Cytokinesis	

5. At which stage do chromosomes "switch" from being composed of two chromatids to one chromatid?

6. a. Why do plants have to do a unique variation of cytokinesis?

   b. Draw a plant cell at the end of cell division. Label the cell plate and explain why it is different from animal cells.

7. Draw a prokaryotic cell. How does bacterial cell division differ from that of animal cells?

Cool Fact: My DNA Is How Long?

DNA molecules are not very wide, only about 3 billionths of a meter across, but they are enormously long. If you stretched out all of the DNA in the 46 chromosomes of the human genome and placed them end to end, the line of DNA would be almost 2 meters long (about 6 feet). So how does all of that DNA fit into a single cell, which is about 25×10^{-6} m across?

Chromatin is the answer. A family of proteins called histones are the major components of the protein complexes that bind to DNA, making the DNA-protein complex known as chromatin. The histones (there are five of them) form a complex themselves that is roughly spherical. DNA wraps around the histone complex like thread on a spool, and each complex can bind about 200 base pairs of DNA. A single DNA molecule (which can be millions of base pairs long) can wrap around the histones, forming a structure that looks like beads on a string. This structure can continue to associate with other proteins and wrap and twist itself into more compact structures. The most compact DNA structures occur at metaphase. So, how does all of that DNA fit inside a cell? Very efficiently!

Chapter 9

TESTING, TESTING

Create a web: Plant cell, animal cell, bacterial cell, cytokinesis, cell wall, plasma membrane, binary fission

Make a flowchart: The cell cycle; explain what occurs in each stage.

Vocabulary Review

Place each term from the following list into the appropriate column of the table below. Some terms may belong in more than one column.

adenine
cleavage furrow
genome
binary fission
cytokinesis
guanine
centrosome
cytosine
microtubules
chromatid
DNA
mitotic spindle
thymine

DNA	Cell Division

PRACTICE TEST

Once you have finished studying this chapter, close your books, grab a pencil, and spend the next 15 to 20 minutes completing this practice test

Compare and Contrast

For each of the following paired terms, write a sentence of comparison ("Both . . . ") and a sentence of contrast ("However, . . . ").

gene/genome

chromosomes/chromatin

autosome/sex chromosome

binary fission/mitosis

Short Answer

1. How does the formation of cell walls in plants differ from that in bacteria?
2. What does DNA code for?
3. Are the X and Y chromosomes homologous? Explain.

4. How are chromosomes evenly divided during mitosis?

Matching

Match the following terms with their description. Each choice may be used once, more than once, or not at all.

a. cytokinesis
b. chromosome
c. chromatin
d. centrosome
e. genome
____ Combination of DNA and protein
____ The complete collection of genetic information
____ The splitting of one cell into two
____ The spindle organizing structure during cell division
____ Individual packets of DNA

Multiple Choice

Circle the letter that best answers the question.

1. All of the following are DNA nucleotides except
   a. adenine
   b. cytosine
   c. guanine
   d. alanine
   e. thymine

2. Which of the following is the last phase of mitosis?
   a. anaphase
   b. telophase
   c. prophase
   d. metaphase
   e. none of the above

3. If a cell has 30 chromosomes before mitosis, how many does each daughter cell have after?
   a. 10
   b. 15
   c. 20
   d. 25
   e. 30

4. The sequence of DNA specifies:
   a. nuclear structure
   b. ribosome activity

c. amino acid sequence of proteins
d. a karyotype
e. none of the above

5. A human cell has_____ chromosomes.

a. 22
b. 23
c. 44
d. 46
e. 48

6. Human males and females have different

a. sex chromosomes.
b. numbers of chromosomes.
c. karyotypes.
d. a and c
e. autosomes.

7. At which stage of the cell cycle does DNA replication occur?

a. G_1
b. S
c. G_2
d. M
e. a, b, and c

8. Plant cells divide by

a. cytokinesis without mitosis.
b. mitosis without cytokinesis.
c. binary fission.
d. binary fusion.
e. mitosis followed by cell-plate formation.

9. Which stage of division is associated with the action of the contractile ring?

a. plant-cell cytokinesis
b. animal-cell anaphase
c. bacterial-cell cytokinesis
d. animal-cell cytokinesis
e. bacterial-cell binary fission

10. Why does mitosis precede cytokinesis?

a. because prokaryotic cells require mitotic products to fuel cytokinesis
b. because DNA takes up most of the inner cell space

c. because it is more important to divide the outside of the cell before the inside
d. because it is more important to divide the genetic material evenly before the cell infrastructure
e. because most biological processes occur in reverse alphabetical order

You decide to give your memoirs to your two daughters. To ensure that each daughter receives an intact set, you copy each page and then file the original and the copy together, connected by a paper clip. When this task is finished, you systematically separate the files into two piles, thereby assuring that each daughter receives one (and only one) copy of each page.

11. In this example, what is analogous to DNA?

    a. the daughters
    b. the paper clips
    c. the cells
    d. the memoirs
    e. the spaghetti

12. In this example, what is analogous to the mitotic stage known as anaphase?

    a. deciding to give the memoirs to your daughters
    b. copying each page
    c. connecting the pages with paper clips
    d. systematically separating the files into two piles
    e. a and c

BUT, WHAT'S IT ALL ABOUT?

Beginning with this unit, we are going to start providing you with fewer and fewer aids to guide you in answering these questions. The answer key will still provide direction about what kind of question is being asked and the type of evidence appropriate to answer the question, but even that will start to become less detailed. Why? Because we are confident that after doing these questions for the first eight chapters, you don't need any more than a nudge in the right direction.

The key experiments that revealed that DNA indeed contains the "recipes" for all cellular components took place only 50 to 60 years ago. Until those classic experiments, many scientists thought that proteins might be the carrier of genetic information. What would be the advantages and disadvantages of proteins as information storage molecules?

1. **What type of question is this?**
   Compare and contrast? Defend a position? Describe the effect of "A" on "B"?

2. **What kind of evidence do you need?**
   This depends in part upon what type of question you think it is. However, since DNA and protein both appear in the question, maybe we need to think about the structures of these molecules and how structure affects function.

3. **Pull it all Together.**
   The question asks for "advantages and disadvantages" – that request should shape your answer.

10

CHAPTER 10

Preparing for Sexual Reproduction: Meiosis

Basic Concepts

- In meiosis, chromosome duplication is followed by two division steps—producing four haploid cells, each with a single copy of each chromosome.
- Meiosis ensures genetic diversity through recombination of homologous chromosome pairs and the independent assortment of chromosomes during meiosis I.
- Meiosis reduces the number of chromosomes by half, making sexual reproduction possible, although not all organisms reproduce sexually.

Steps in Meiosis

- Meiosis reduces a diploid parental cell into four haploid daughter cells because the single round of DNA duplication is followed by two rounds of cell division, each of which divides the DNA equally between the daughter cells.
- The steps of meiosis I involve matching and separation—chromosome pairs are aligned along the metaphase plate, and then one of each chromosome pair is randomly assorted into each of the two daughter cells.
- Meiosis II consists of cell division without DNA duplication, the sister chromatids of each chromosome are pulled apart during anaphase II into the daughter cells (similar to division during mitosis).

THINK ABOUT IT

1. Use the following table to place the listed steps of meiosis in the correct order and describe the key events occurring at each stage: prophase II, anaphase I, metaphase II, telophase I, telophase/cytokinesis II, prophase I, metaphase I, anaphase II.

Stage	Key Events

2. Describe the genetic content and chromosome structure of cells at the end of meiosis I.

3. Describe the products of meiosis and chromosome structure in meiotic products.

4. Which stage of meiosis most closely resembles mitosis? Why?

Significance of Meiosis

- Meiosis produces gametes, haploid cells that can fuse to make a diploid zygote, thus avoiding the problem of genome duplication each time sexual reproduction occurs.
- Meiosis increases genetic diversity. Chromosome pairs can intertwine and exchange pieces when pairing during prophase I. Also, because chromosome pairs align along the equatorial plane independently during anaphase I, maternally and paternally derived chromosomes get mixed within the daughter cells.

THINK ABOUT IT

1. A newly discovered, sexually reproducing plant has just been discovered in the Amazon rain forest. There are 32 chromosomes in most of its cells.

   a. What is its diploid chromosome number?

   b. How many chromosomes would you expect to find in its gametes?

2. Explain why meiosis increases genetic diversity. Describe an example from your own family.

3. Draw two possible ways for two pairs of homologous chromosomes to align on the metaphase plate, and generate four unique gametes. Use different colors for the homologous chromosomes and sister chromatids.

Reproduction: Human and Otherwise

- All sexually reproducing organisms exist in two sexes and produce gametes that fuse to make a new organism.
- In humans, the X and Y chromosomes segregate during meiosis, producing gametes that bear either an X chromosome or a Y chromosome. Sex is determined when the zygote forms—fusing two X-bearing gametes produces the human female, and fusing one X-bearing gamete and one Y-bearing gamete produces the human male. Fusion of two Y-bearing gametes is not possible.

- Males produce sperm in the testes from precursor cells (spermatogonia). Sperm are an efficient delivery system for 22 autosomes and either an X or Y chromosome.
- Females produce eggs from precursor cells (oogonia) with sufficient cellular content to support the growth of the zygote produced following fusion with a sperm. Each egg contains 22 autosomes and an X chromosome.
- Some organisms reproduce asexually, producing offspring from the parent through a mitotic process—binary fusion in bacteria and vegetative reproduction in plants.

THINK ABOUT IT

1. Draw a mature sperm cell. Which organelle provides the energy for sperm propulsion? Which part of the sperm serves as the propulsion mechanism?

2. Use the following table to compare and contrast spermatogenesis and oogenesis in humans.

Feature	Spermatogenesis	Oogenesis
Gametes produced		
Chromosomes per gamete		
Arrest at any stage?		
Gamete like a functioning cell?		
Equal cytokinesis?		

3. Which generates more diversity—sexual or asexual reproduction? Why?

4. What is an advantage of asexual reproduction?

Cool Fact: Chromosomal Sex Determination

The human "odd couple" pair of chromosomes is responsible for determining whether you spend your mature life hunting for the perfect pair of black pumps or refusing to ask directions. In humans, the female sex results when two X chromosomes are present in the zygote, whereas zygotes bearing one X and one Y chromosome become male. The Y chromosome carries genes that deliver the message, "develop male sexual characteristics" to the embryo, and it is essential for maleness.

Other species have dealt differently with the problem of how to direct the development of two sexes—many don't require entire chromosomes for sex determination, just a sex-determination gene. Insects, for example, use a haplo-diploid mechanism. Males develop when either the chromosome pair has only one allele (hemizygous) or two of the same alleles (homozygous). Females are heterozygous; they have two different alleles on their pair of chromosomes. Birds have a W-Z system for sex determination. In this system, the W gene sends the signal to become female, so female birds are WZ and males are ZZ genotype.

TESTING, TESTING

Create a web: meiosis, mitosis, daughter cell, haploid, diploid, life cycle, cell cycle

Make a flowchart: Section 10.2, "The Steps in Meiosis"

Vocabulary Review

Place each term from the following list into the appropriate column of the table below. Some terms may belong in more than one column.

anaphase oogonium prophase spermatid

metaphase polar bodies recombination telophase tetrad

Meiosis	Gametogenesis

PRACTICE TEST

Once you've finished studying this chapter, close your books, grab a pencil, and spend 15 to 20 minutes working on this practice test.

Compare and Contrast

For each of the following paired terms, write a sentence of comparison ("Both . . .") and a sentence of contrast ("However, . . .").

chromosome/chromatid

haploid/diploid

meiosis/mitosis

spermatocytes/oogonia

Matching

Match the following terms with their description. Each choice may be used once, more than once, or not at all.

a. resembles mitosis
b. sperm production
c. homologous chromosomes pair up
d. haploid structures that do not complete gamete formation
e. process by which genetic material is exchanged between homologous chromosomes

___ Recombination
___ Polar bodies
___ Spermatogenesis
___ Prophase I
___ Meiosis II

Short Answer

1. What is the significance of meiosis?

2. How common is meiosis when compared to mitosis?

3. In humans, males can produce many more sperm than females can produce eggs. Explain.

4. What is the major difference between offspring produced sexually and those produced asexually?

Multiple Choice

Circle the letter that best answers the question.

1. Which of the following is true regarding sperm?

   a. They have all the typical organelles.
   b. One is produced for every spermatogonium.
   c. They are haploid.
   d. They are much larger than human eggs.
   e. They are somatic cells.

2. The sex of a human fetus is determined by

   a. the age of the mother.
   b. the age of the father.

c. the phase of the moon at the time of conception.
d. the sex chromosome found in the sperm.
e. the sex chromosome found in the egg.

3. In humans, meiosis produces cells that are

a. identical.
b. haploid.
c. gametes.
d. all of the above.
e. b and c only

4. The process of growing a plant cutting into a plant is an example of

a. asexual reproduction.
b. sexual reproduction.
c. meiosis.
d. oogenesis.
e. spermatogenesis.

5. During meiosis, homologous chromosomes separate at

a. prophase I.
b. metaphase I.
c. anaphase I.
d. metaphase II.
e. anaphase II.

6. Which of the following allows sexual reproduction to occur generation after generation?

a. mitosis.
b. mucosis.
c. meiosis.
d. cytosis.
e. halitosis.

7. Recombination occurs

a. during mitosis I.
b. during anaphase II.
c. during prophase I.
d. during fertilization.
e. during binary fission.

8. Why are no two gametes exactly alike?

a. because their cell membranes are unique, like snowflakes.
b. because each gamete undergoes a slightly different version of mitosis during its formation.

c. because each gamete has a different combination of parental chromosomes.
d. Both b and c are correct.
e. All of the above are correct.

9. In meiosis II,

a. chromatids are separated and sent to separate daughter cells.
b. homologous chromosomes are separated into different daughter cells.
c. haploid chromosomes fuse to make a diploid cell.
d. bacteria make exact copies of their chromosomes.
e. all of the above are correct.

10. Meiosis is

a. shared by all living organisms.
b. found only among bacteria.
c. found only in sexually reproducing organisms.
d. found in every cell of the body.
e. a seven-letter word for sushi.

11. Reginald Prettiboy decides to clone himself. What sex would the "offspring" be?

a. male
b. female
c. hermaphrodite
d. either male or female
e. asexual

12. Which of the following is NOT common to both oogenesis and spermatogenesis?

a. They are both formed by two rounds of meiosis.
b. They both begin with a diploid cell.
c. They both produce gametes.
d. They both result in the formation of four haploid gametes.

13. You discover a new type of animal that you have named *Quinntella georgiana.* Georgiana is a type of sea star, which means that it is capable of reproduction by

a. regeneration.
b. hermaphroditisim.
c. binary fission.
d. vegetative reproduction.

BUT, WHAT'S IT ALL ABOUT?

Here's a question to help you pull together what you've learned so far using this text. If you don't remember how to attack this type of question, check out the example in Chapter 1.

Meiosis eliminates one problem of sexual reproduction by reducing by half the number of chromosomes in the gametes. Sometimes, however, this dance of the chromosomes goes awry when the chromosomes are not divided evenly among the four resulting gametes. In Chapter 12, we will consider in more detail what happens when gametes end up with too many or too few chromosomes, but for now let's think about how this error might happen. Think about how the process of meiosis proceeds when making human sperm and oocytes; when or how do you think it is most likely that chromosomes will not segregate properly? To help you with your thinking, consider the observation that missegregation of chromosomes happens more often in older humans.

1. **What type of question is this?**
   Compare and contrast? Defend a position? Describe the effect of "A" on "B"?

2. **What kind of evidence do you need?**
   What are the key parts of this question: Chromosomal missegregation in meiosis, and the process of meiosis in human gamete formation? This question appears to be limited to the information you found in this chapter.

3. **Pull it all Together.**
   You can do it! And this question is probably a lot harder than you would find on an exam at this stage of the course, so it is really good practice for your critical thinking skills.

CHAPTER 11

The First Geneticist: Mendel and His Discoveries

Basic Concepts

- The physical appearance of an organism, the phenotype, is determined by the function and interactions of the proteins encoded by the organism's genome (genotype).
- Gregor Mendel's quantitative studies on the inheritance of physical traits in pea plants first revealed three basic laws of genetics:
  —Genetic elements come in pairs.
  —Genetic elements separate when gametes are made.
  —Genetic elements segregate independently of each other.
- Variations in phenotype caused by protein interactions and environmental effects can still be accounted for by Mendel's laws.

Phenotype and Genotype

- Phenotype refers to the visible characteristics of an organism, its actions, function, size, shape, and so on. The proteins encoded by an organism's genes determine these observable traits.
- Mendel's achievement was to infer principles that explain the heritability of traits by matching inputs with outputs without knowing anything about the physical mechanism involved in genetic information transfer.
- Mendel's experimental organism, *Pisum sativum*, has seven discernible traits; seed color and shape, pod color and shape, flower color and location, and plant height. Plants can be breed true for each trait, so the "inputs" can be determined with absolute certainty.

THINK ABOUT IT

1. What were the three main inferences that Mendel made?

2. What field of biology is Mendel considered the "father" of?

3. What is the relationship (in human terms) of peas in a pod?

4. What type of mating can pea plants do that humans can't?

5. A mutation in a chloride channel gene is responsible for the recessive genetic disorder, cystic fibrosis, in which the lungs are filled with a thick mucus that impairs their function. How would you describe the genotype and phenotype of a person with cystic fibrosis?

Mendel's Studies

- By counting how often a trait appears among the offspring of a cross of true-breeding, parental plants, Mendel quantified the frequency of each trait within the filial generation.
- From these proportions, Mendel induced that
  —Traits come in pairs, and that one trait of the pair "dominates" over the other.
  —Traits do not "blend" (show an intermediate phenotype), because the "recessive" trait can always be recovered when plants of the first filial (F1) generation are crossed.
  —The genetic elements responsible for each trait, which we now refer to as the gene responsible for each phenotype, segregate into the gametes independently of each other.

THINK ABOUT IT

1. Consider a family. There are grandparents, their children, and then their grandchildren. Which would be the P, F_1, and F_2?

2. You find a new plant species, some having orange flowers and some having yellow flowers. When you cross a yellow- and an orange-flowered plant, all the offspring have orange flowers. Which flower color is dominant? If flower color behaves the same way as pea color, what ratio of yellow- and orange-flowered plants do you expect in the F_2?

3. Which of the following would be alleles of the same gene?

   a. yellow peas b. blue eyes c. left foot

   d. brown eyes e. ears f. green peas

4. Which stage of meiosis is responsible for Mendel's Law of Segregation? What actually segregates?

5. Is a yellow-flowered plant that produces nothing but yellow-flowered plants considered to be homozygous or heterozygous for flower color?

6. You have two purple plants, which you cross to green plants:

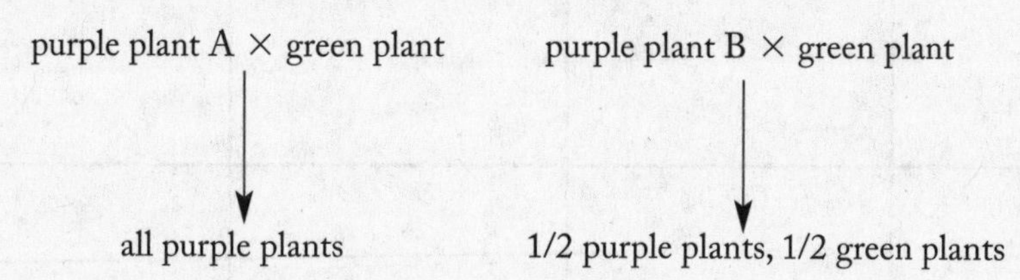

a. Which color (purple or green) is dominant?

b. Which of the two purple plants (A or B) is homozygous and which is heterozygous?

7. a. Fill in Punnett square 1 for the cross *Aa* × *Aa*.

Square 1

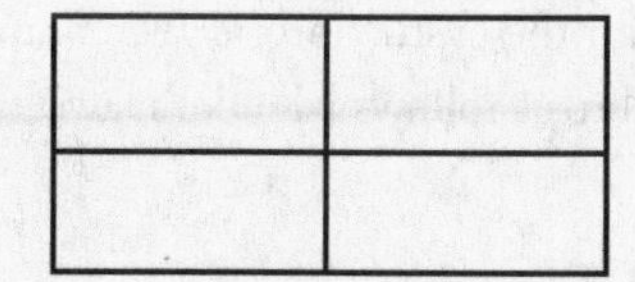

b. What are the genotypes of the parents in square 2?

Square 2

BB	*Bb*
Bb	*bb*

c. Continue filling in square 3:

Square 3

	AB	*Ab*	*aB*	*ab*
AB				
Ab		*Aabb*		
aB			*aaBb*	
ab	*AaBb*			

d. List all the genotypes from square 3 in the appropriate columns below:

Phenotype	*A_B_*	*A_bb*	*aaB_*	*aabb*
Genotypes				
Total genotypes				

e. Do your totals agree with Mendel's dihybrid phenotypic ratios? Why or why not?

8. Which stage of meiosis is responsible for Mendel's Law of Independent Assortment?

9. Two parents, both of whom are heterozygous for the cystic fibrosis gene mutation, have four children. The inheritance of cystic fibrosis follows Mendel's laws.

a. Draw the Punnett square for these parents.

b. On average, how many children (out of four) would you expect to be homozygous for the cystic fibrosis gene mutation?

Variations on Mendel's Insights

- Incomplete dominance of one trait over another occurs when one allele of the gene responsible for the phenotype is nonfunctional—it does not produce a product, as in the snapdragon example where one "white flower" allele does not produce a pigment protein. The phenotype of the heterozygous organism seems to be intermediate between the parental types, but it simply reflects the presence of half of the concentration of functional protein.
- Codominant traits result when different alleles of a single gene each produce functional proteins. As in the A and B blood types, when both alleles are present, both proteins are made and neither protein "dominates" over the other.
- Single gene/single phenotype occurs rarely; most gene products interact with each other, producing multiple phenotypes.
- Gene expression can be affected by environmental conditions, producing variable phenotypes among genetically identical organisms.

THINK ABOUT IT

1. In a plant, orange flowers are dominant to white flowers. Draw a Punnett square for a cross between two plants with orange flowers (*Oo*). Highlight the boxes using distinct colors and shadings. What is the phenotypic ratio? What is the genotypic ratio? Which of these reflects the phenotypic ratio that would be observed in cases of incomplete dominance?

2. In the preceding question, how do different alleles of the color genes cause differences in flower color?

3. Which blood type best illustrates the concept of codominance?

4. List the multiple alleles at the ABO blood type gene.

5. Individuals with Marfan syndrome experience several health problems, including cardiac, eye, and joint problems. What is an explanation for this?

6. Why can a single gene have so many diverse effects?

7. Not every woman with the "breast cancer" gene mutation will develop breast cancer. Why might this be?

8. Can you think of any other examples of interactions of genes with the environment?

Cool Fact: Mendel and Mitochondrial Inheritance

Mendel's landmark studies justify his distinction as the "father of genetics," but his choice of experimental organism was a major part of his success. Fortunately, the seven traits of the garden pea are encoded by single genes, each located on one of the pea's seven chromosomes. Mendel's proportions were not complicated by other factors affecting the sorting of genes during meiosis. This relatively simple system allowed Mendel to quantify the principles against which all later observations would be measured, providing a theory to explain other factors affecting the inheritance of a phenotype.

Some phenotypes were particularly troublesome to explain within Mendel's laws, such as maturity onset diabetes of the young (MODY). MODY follows a pattern of maternal inheritance that is not sex-linked; that is, all of the children of a MODY mother inherit the MODY phenotype. But the phenotype is highly variable in severity, so some of these offspring may seem not to be affected by MODY even though they carry the MODY genotype.

MODY's inheritance pattern was explained once it was determined that the genes responsible for MODY were not in the nucleus but in the mitochondria. Shortly after fertilization the sperm mitochondria that have entered the oocyte are destroyed, so the zygote possesses only maternally inherited mitochondria. These maternal mitochondria replicate as the embryo grows, and they populate virtually all the cells of the growing fetus. However, mitochondrial numbers among cells are quite variable. If only some of the mitochondria in the oocyte carry the MODY genotype, then some of the fetus' cells may have no MODY-carrying mitochondria, some will have a few, and some will have many, depending on how the original mitochondrial population replicates and assorts itself. This random and variable distribution of MODY mitochondria explains the variation in onset and severity of diabetes in affected people.

Despite the "odd" inheritance pattern of the MODY phenotype, the genotype inheritance pattern still follows Mendel's laws—as Mendel would probably have predicted.

Chapter 11

TESTING, TESTING

Create a web: dominant, recessive, allele, homozygous, heterozygous, phenotype, genotype, trait

Make a flowchart: Follow a given trait from a parental autosome cell through meiosis into a gamete into the embryonic autosome. At each stage, decide whether the cell is haploid or diploid, homozygous or heterozygous.

Vocabulary Review

Place each term from the following list into the appropriate column of the table below. Some terms may belong in more than one column.

	heterozygous	
allele	incompletely dominant	trait
codominant	pea color	
dominant	recessive	

Phenotype	Genotype	Inheritance Pattern

PRACTICE TEST

Once you have finished studying this chapter, close your books, grab a pencil, and spend the next 15 to 20 minutes taking this practice test.

Compare and Contrast

For each of the following paired terms, write a sentence of comparison ("Both . . .") and a sentence of contrast ("However, . . . ").

genotype/phenotype

allele/gene

Law of Segregation/Law of Independent Assortment

dominant/recessive

Matching

Match the following terms with their description. Each choice may be used once, more than once, or not at all.

a. alleles
b. B genes
c. phenotype
d. pleiotropy
e. codominance

—Physical appearance of an organism
—When alleles cannot mask or "cover" each other
—Different versions of the same gene
—Specify instructions for building proteins
—When one gene affects multiple aspects of the phenotype

Short Answer

1. Explain why some traits may appear to follow a non-Mendelian? inheritance pattern.

2. Explain Mendel's first and second laws in relation to meiosis.

3. What is a gene?

4. Explain the relationship between environment and phenotype.

Multiple Choice

Circle the letter that best answers the question.

1. Which of the following was used by Mendel to prove his hypotheses?

   a. mice
   b. snapdragons
   c. peas
   d. chickens
   e. maple trees

2. You perform a cross between red yucky bugs and white yucky bugs and see all pink yucky bugs. What is the relationship between the red and white yucky bug color?

   a. Red is dominant to white.
   b. Red is incompletely dominant.
   c. White is dominant to red.
   d. Red is pleiotropic.
   e. None of the above.

3. You carry out a selfing of the F1 of a yellow and green pea monohybrid cross. Which of the following phenotypic ratios do you expect to see in the F2?

   a. 9:3:1
   b. 1:2:1
   c. 3:1
   d. 9:3:3:1
   e. 9:3:4

4. What do the terms gene and allele have in common?

   a. Both refer to the type of pea plant Mendel developed.
   b. Both could refer to the same DNA sequence.
   c. Both refer to famous baseball players of the 1950s.
   d. Both refer to the blending of inherited traits.
   e. They have nothing in common.

5. A single allele of a gene that exerts multiple phenotypic effects is due to

   a. multiple alleles.
   b. interactions with the environment.
   c. pleiotropy.
   d. independent assortment.
   e. segregation.

6. In a monohybrid cross, a 3:1 phenotypic ratio is observed in the F_1. What is the underlying genotypic ratio?

   a. 3:1
   b. 9:3:3:1

c. 1:1
d. 1:2:1
e. 1:2

7. In order for a person to have cystic fibrosis, they must be homozygous recessive for the disease allele. If two heterozygous parents have many children, what is the expected ratio of those with cystic fibrosis to those without cystic fibrosis?

   a. 3 with:1 without
   b. 1 with:3 without
   c. 4 with:1 without
   d. 1 with:4 without
   e. None of the above

8. The major difference between dominant and recessive alleles of the same trait is that

   a. when both are present, only the recessive alleles are expressed.
   b. when both are present, neither is expressed.
   c. when both are present, only the dominant is expressed.
   d. when neither is present, both are expressed.
   e. expression is in the eye of the beholder.

9. Which of the following is not a possible offspring type from the mating of *AA* and *Aa*?

   a. *AA*
   b. *Aa*
   c. *aa*
   d. All of the above are possible.
   e. None of the above are possible.

10. When Mendel followed the inheritance of two traits at the same time, he found

    a. The inheritance pattern of one trait would always control the other.
    b. No conclusions could be drawn from his results.
    c. The traits passed from one generation to the next independently of each other.
    d. One trait increased the expression of the other.
    e. The lost mines of Kilimanjaro.

11. Reginald Pisum is a homozygous dominant for the trait Sativum, or pea-like ear-lobes. If he were to marry Pearl Finkle, who is heterozygous for the trait, what proportion of their offspring will also be homozygous dominant?

    a. 0
    b. 0.25
    c. 0.50

d. 0.75
e. 1.00

12. Reginald Pisum is homozygous dominant for the trait Sativum. How did he inherit his alleles?
    a. both dominants from his mother
    b. both dominants from his father
    c. dominant from his mother, dominant from his father
    d. dominant from his mother, recessive from his father
    e. dominant from his father, recessive from his mother

13. What is the phenotypic ratio for the offspring of a cross between a homozygous dominant (ZZ) and a homozygous recessive (zz)?
    a. All dominant
    b. All recessive
    c. Half dominant, half recessive
    d. 1 dominant, 1 heterozygous, 2 recessive
    e. 1 dominant, 2 heterozygous, 1 recessive

BUT, WHAT'S IT ALL ABOUT?

Here's a question to help you pull together what you've learned so far using this text. If you don't remember how to attack this type of question, check out the example in Chapter 1.

The hallmark of Mendel's work was his careful record keeping and large sample size; his quantitative results lead to identifying the phenotypic ratios that he used to infer the laws of independent assortment and segregation. Mendel was lucky, too, to have picked 7 phenotypes that were all independent. What would have happened to his analysis if 2 of the 7 traits were not independent, say if all yellow seeds were always smooth and all green seeds were always round?

1. **What type of question is this?**
   Compare and contrast? Defend a position? Describe the effect of "A" on "B"?

2. **What kind of evidence do you need?**
   What did Mendel's ratios tell him about the inheritance of traits? If 5 of the traits produced predictable ratios leading to the description of "dominant" and "recessive" traits, and 2 did not, how do you think Mendel would have interpreted it, given his knowledge base?

3. **Pull it all together.**
   Go for it!

CHAPTER 12

Chromosomes and Inheritance

Basic Concepts

- Human sex chromosomes play a unique role in heredity.
- Malfunctioning chromosomes change the phenotype of the organism, indicating that chromosomal makeup and function are critical to human health.
- Phenotype changes may be caused either by defective alleles, by physical damage to the chromosome, or by incorrect separation of chromosomes during gametogenesis.

Sex-Linked Inheritance in Humans

- Defects in alleles on the X chromosome are expressed more often in males than in females because males have a single X chromosome; consequently, color blindness and hemophilia (diseases caused by defective X-chromosome genes) occur more often in men than in women.
- An "extra" or a missing sex chromosome causes pleiotropic effects, disrupting development of the embryo.

THINK ABOUT IT

1. Why are more males than females affected by disease genes on the X chromosome?

2. Your friend Becky tells you of a strange skin condition found in her family. Men on her father's side share a strange, camouflage-like melanin pattern on their backs. The condition is never seen in females of this lineage.

   a. Could this trait be genetic?

   Why?

   b. Could it be sex linked?

   Why?

Malfunctioning Chromosomes

- Changes in gene sequences carried upon the autosomes can also produce nonfunctional or poorly functioning proteins, causing human disease. These autosomal genetic diseases can be detected by examining family pedigrees.
- Diseases caused by nonfunctional alleles may be recessive, requiring the presence of two defective alleles to produce the disease phenotype, or dominant, producing the disease phenotype even when one functional allele is present.
- Disease can be caused by having more or fewer chromosomes than a normal diploid number, as in Down syndrome. This aneuploidy occurs when the chromosomes segregate improperly during meiosis.

THINK ABOUT IT

1. If you were a carrier for sickle-cell anemia, would you produce more malaria-safe children by marrying a normal (homozygous) person, or another carrier? Draw the two Punnett squares.

2. Marfan syndrome is an autosomal dominant disorder. If a child has Marfan syndrome, is it likely that either of his or her parents also has it?

3. What is the likely fate of a human embryo with three sets of 23 chromosomes?

4. For each of the following terms, write a P and/or A to symbolize whether it would be found in plants, animals, or both:

   ___ diploid ___ haploid ___ polyploidy

5. Circle the chromosome number(s) that would represent aneuploidy in a human.

   45 46 47

6. Draw a meiosis I and a meiosis II nondisjunction of chromosome 21. Use different colors for the homologous chromosomes and sister chromatids.

7. What condition does a nondisjunction of chromosome 21 lead to?

8. What is an example of a sex-chromosome abnormality?

Structural Aberrations in Chromosomes

- Deletions occur when pieces of chromosomes are lost during meiotic recombination. Large-scale deletions, such as loss of parts of chromosomes, are easily detected in karyotype spreads.

- Inversions and translocations, in which part of a chromosome is "flipped" or exchanged with part of another chromosome, are also detectable in karyotype spreads.
- Duplications occur when meiotic crossing over between homologous chromosomes is unequal. The result is that one chromosome of the pair loses sequences, while the other gains a duplicate sequence.

THINK ABOUT IT

1. Use this chromosome to

A B C D E

a. Draw a deletion of A.

b. Draw an inversion between B and D.

c. Draw a duplication of C.

Cool Fact: Gene Therapy

Inherited diseases are considered incurable because a cure would require replacing the defective gene with a functional copy of the gene in every cell in the individual—a feat not possible with our current technology. But what if doctors could introduce a functional gene into enough cells in specific tissues to enable production of the functional protein where it is most needed? Even if the gene was not part of the genome, gene function could be restored and the patient could be "cured."

Such is the concept behind gene therapy. Using recombinant DNA cloning techniques (Chapter 15), a copy of a gene can be introduced into a modified version of a human virus. The virus carries the gene into tissues, where it can specify the synthesis of the missing protein and restore function. Human trials of gene therapy have been carried out to test its effectiveness for several autosomal genetic diseases. The results of these trials have generally been less successful than hoped for. Even when the gene does manage to enter the appropriate tissue and start to produce the functional protein, the effect doesn't last. The most serious problem, however, is that most of these vectors cause the body to mount an immune response against the vector. This means that if the injection is repeated, the body eliminates the vector just as it would any foreign invader—and the gene never makes it to its target. The extreme consequence of this immune response against the gene-therapy vector can be death, as was the case for Jesse Gelsinger, a young man who died as a result of a gene therapy trial for the treatment of ornithine transcarbamolyase deficiency.

Despite these setbacks, researchers remain confident that the immunological problems will be overcome. Gene therapy will be a powerful addition to our medical arsenal, and not just for autosomal genetic diseases. Scientists are also studying how to use gene therapy to introduce genes that will affect specific functions (not a specific gene) within a cell. This approach, which promises new treatment for diseases such as cancer, arthritis, and heart disease, is an approach that will deliver therapeutic agents exactly where they are needed.

Chapter 12

TESTING, TESTING

Create a web: allele, nondisjunction, aneuploidy, polyploidy, diploid, haploid

flowchart: Demonstrate the inheritance of a Y-linked trait. Can daughters be carriers?

Vocabulary Review

Place each term from the following list into the appropriate column of the table below. Some terms may belong in more than one column.

aneuploidy | hemophilia | red-green color blindness
cri-du-chat | Huntington | sickle-cell anemia
Down syndrome | Klinefelter | Turner

Phenotype	Genotype

PRACTICE TEST

After you have finished studying this chapter, close your books, grab a pencil, and spend the next 15 to 20 minutes working on this practice test.

Compare and Contrast

For each of the following paired terms, write a sentence of comparison ("Both . . . ") and a sentence of contrast ("However, . . . ").

aneuploidy/polyploidy

inversion/translocation

Turner/Klinefelter

sex-linked recessive/autosomal recessive

Matching

Match the following terms with their description. Each choice may be used once, more than once, or not at all.

a. inversion
b. translocation
c. deletion
d. duplication
e. nondisjunction

___ Chromosomes fail to sort properly during meiosis
___ Exchange of material between non-homologous chromosomes
___ A segment of the sequence is turned over and re-inserted
___ Certain sequences are copied more often than others
___ A portion of the message is completely lost

Short Answer

1. What is different about the inheritance of traits located on the sex chromosomes?

2. Suppose that a deadly autosomal recessive disorder were to suddenly become dominant. What changes would you predict?

3. What is the difference between a translocation and a non-disjunction?

4. How common is polyploidy in humans?

Multiple Choice

1. An individual with an autosomal recessive disorder generally has ___ copies of the dominant allele.

   a. 0
   b. 1
   c. 2
   d. 2, on the X chromosome
   e. 2, on the Y chromosome

2. Which of the following is an autosomal disease?

   a. sickle-cell anemia
   b. red-green color blindness
   c. hemophilia
   d. Turner syndrome
   e. Klinefelter syndrome

3. X-linked disorders:

   a. are more common in females than males
   b. never occur
   c. are more common in males than females
   d. occur equally frequently in males and females
   e. affect the Y chromosome

4. An older mother may be at increased risk for having a child with which of the following disorders?

   a. Huntington disease
   b. Turner syndrome
   c. Down syndrome
   d. Klinefelter syndrome
   e. None of the above.

5. Some flies have a diploid chromosome number of 8. Which of the following chromosome counts represents an aneuploid chromosome number?

   a. 0
   b. 4
   c. 8
   d. 9
   e. 16

6. Which of the following chromosome numbers would be found in a polyploid human embryo?

   a. 23
   b. 46
   c. 69
   d. 47
   e. None of the above.

7. Which of the following is a syndrome that is more likely to be found in Reginald Pisum than his wife Pearl?

   a. Turner syndrome
   b. Klinefelter syndrome
   c. color blindness

d. both a and c
e. both b and c

8. Which of the following describes a situation in which the gene is not lost, but simply misplaced?

a. translocation
b. transcription
c. transportation
d. transcendental
e. transformation

9. A new student joins your class. She is small of stature and does not seem to have developed sexually. You suspect the she might have

a. red-green color blindness
b. hemophilia
c. sickle-cell anemia
d. Turner syndrome
e. Huntington disease

10. Which of the following is not a change in the structure of a chromosome?

a. deletion
b. translocation
c. inversion
d. non-disjunction
e. duplication

11. A mutation occurs in a gene that codes for black coat color in a rare species known as "Crying" hyenas. Which of the following phenotypes is most likely to be the result of such a change?

a. baldness
b. four ears
c. white coat
d. no tail

12. Three rat embryos are generated during an *in vitro* fertilization experiment. Of the genotypes listed below, which would be male?

a. XO
b. XXY
c. XXX
d. A and B
e. None of these.

BUT, WHAT'S IT ALL ABOUT?

Here's a question to help you pull together what you've learned so far using this text. If you don't remember how to attack this type of question, check out the example in Chapter 1.

Identical twins result when a single, fertilized egg splits into two cells during the earliest stages of development and each individual cell develops into an embryo. Despite a common set of genes, these twins still do not mature into indistinguishable people – although physically, mentally, and emotionally very similar, they are clearly different people. If the genotype drives the phenotype, why aren't identical twins identical people?

1. **What type of question is this?**
   Compare and contrast? Defend a position? Describe the effect of "A" on "B"?

2. **What kind of evidence do you need?**
   What kind of information do we need to provide to explain how genetically identical people can be different?

3. **Pull it all together.**
   Go for it!

CHAPTER 13

DNA Structure and Replication

Basic Concepts

- The molecular structure of DNA, or the three-dimensional order of the atoms in space, was proposed in 1953 by James Watson and Francis Crick, based on the experimental data of Rosalind Franklin and Maurice Wilkins. Watson and Crick's model revolutionized the study of genetics and ushered in the field of molecular biology.
- The double helix of DNA consists of two repeated polynucleotide chains associated through hydrogen bonds between the nitrogen bases. The complementarity of the bases means that a DNA molecule can be a template for its own replication. The order of the bases along the chain constitutes a code for specifying the order of amino acids within a protein.
- Changes in genetic information, called mutations, occur when the sequence of nucleotide bases in the DNA change.

Watson and Crick

- Determining how DNA could store and transmit genetic information required understanding the physical structure of the molecule.
- Watson and Crick's seminal, double-helix model was inferred using experimental data from several research teams, particularly data from the x-ray crystallography experiments of Rosalind Franklin and Maurice Wilkins.

THINK ABOUT IT

1. How did Watson and Crick elucidate the structure of DNA?

2. Who did Watson and Crick collaborate with to elucidate the structure of DNA?

The Double Helix

- DNA is composed of sugar molecules linked together by phosphate bonds. Each sugar molecule also has one of four nitrogen-containing molecules bound to it—adenine, thymine cytosine, or guanine. These nitrogen bases can form hydrogen bonds with each other—adenine with thymine and cytosine with guanine. This hydrogen bond can hold two DNA chains together such that the hydrogen-bonded nitrogen bases are on the inside of the helix and the sugar-phosphate bonds form the outside edges.

- The specificity of the hydrogen bonding between the bases explains the complementarity of adenine for thymine and cytosine for guanine, and it indicates how each DNA strand can serve as a template for the replication of the helix.
- The order of the bases along the DNA strand serves as a code that specifies how the protein chain is to be assembled.

THINK ABOUT IT

1. If a strand of DNA has the sequence A T C C G A T C and serves as the template for the synthesis of a new strand, write in the sequence of the new strand.

2. What types of enzymes are necessary for the copying of DNA?

3. Is DNA replication completely error free? What measures does a cell have to deal with errors?

4. Fill in the blanks in the correct order using the following terms: sugar, nucleotide, phosphate, DNA, base, gene, chromosome.

Mutations

- Mutation, a permanent change in the sequence of bases in a DNA molecule, may occur when bases are mispaired during replication. Subsequent rounds of mitosis preserve the change and pass it along to daughter cells.
- A mutation in the DNA of the germ-line cells, the gametes, allows the changed DNA to appear in every cell in the organism and passes the change from one generation to the next.
- Mutations are usefully harmful, compromising the function of the protein specified by the DNA. On rare occasion, mutations may add new information to the genome and thus provide the raw material for evolutionary adaptations.

THINK ABOUT IT

1. If a cell acquires a mutation and then undergoes mitosis, will the daughter cells also have the mutation?

2. What is one potential consequence of a mutation in a skin cell?

3. Will a child of a person with a mutation in a skin cell inherit that mutation?

4. Are mutations always harmful?

Cool Technique: Site-Directed Mutagenesis

Mutations are usually harmful to the organism because even a single base change in the DNA sequence may cause major changes to the structure of the protein encoded by that gene. Because a protein's structure determines how well it functions, even small changes can have big effects on the cell's ability to carry out its work.

Scientists can exploit mutations to learn exactly which amino acids within a protein are most important for the protein's function. In site-directed mutagenesis, single base changes are introduced into the DNA sequence that encodes for the protein of interest. Because only one base is changed at a time, only one amino acid is affected. These mutated genes can be introduced into the experimental organism, so that the altered protein's function within the organism can be studied. If the mutation causes the protein to malfunction, scientists can use other techniques to determine where the change occurs on the three-dimensional structure of the protein. Thus it is possible to map a protein's function to specific regions of its physical structure by methodically changing the genetic code.

Chapter 13

TESTING, TESTING

Create a web: nucleotide, phosphate, base, DNA, RNA, protein, adenine, ribose
Make a flowchart: Chart the chronological order of discoveries about DNA structure and function. If possible, try to figure out what was known about DNA when your parents and grandparents were the age you are now.

Vocabulary Review

Place the following terms in the appropriate column of the table.

A-C pair* DNA ligase G-T pair*

A-T pair* DNA polymerase nucleotide sequence

cancer G-C pair* point mutation

*Note: These are new base pairs formed during DNA replication.

DNA Replication	Mutation

PRACTICE TEST

After you have finished studying this chapter, close your books, grab a pencil, and spend the next 15 to 20 minutes taking this practice test.

Compare and Contrast

For each of the following paired terms, write a sentence of comparison ("Both . . . ") and a sentence of contrast ("However, . . . ").

DNA polymerase/DNA ligase

somatic cell mutation/germ-line mutation

phosphate/base

cancer/genetic variability

Matching

Match the following terms with their description. Each choice may be used once, more than once, or not at all.

a. deoxyribose
b. adenine
c. replication
d. mutation
e. polymerase

____A change in the gene sequence
____Process of creating "new" chromosomes
____One of the nitrogen-containing bases
____The sugar portion of a nucleotide
____The assembly and error-checking molecule

Short Answer

1. What three molecules comprise the components of DNA?

2. What forms the handrails (or backbone) of the DNA double helix?

3. What is the function of DNA polymerase?

4. How do carcinogens commonly cause cancer?

Multiple Choice

1. How may mutations arise?

   a. Proofreading
   b. Uncorrected errors during DNA replication

c. Temporary insertion of the wrong nucleotide in a new DNA strand
d. All of the above
e. None of the above

2. What is/are possible consequences of mutations?

a. Faster DNA replication
b. Cancer
c. Evolution
d. Unequal cytokinesis
e. b and c

3. What is required for DNA replication?

a. DNA polymerases
b. DNA ligase
c. a template
d. a and b
e. a, b, and c

4. A mutation in a liver cell

a. will be inherited by the daughter cells of that liver cell.
b. will be passed on to offspring.
c. may lead to liver cancer.
d. a and c
e. None of the above.

5. Where is the bulk of the DNA found in an animal cell?

a. the nucleoid
b. the nucleus
c. the mitochondria
d. ribosomes
e. endoplasmic reticulum

6. Which of the following is in the correct order from smallest to largest?

a. chromosome, gene, nucleotide, phosphate
b. gene, nucleotide, chromosome, phosphate
c. phosphate, chromosome, nucleotide, gene
d. nucleotide, phosphate, gene, chromosome
e. phosphate, nucleotide, gene, chromosome

7. The study of genetics at the level of DNA is known as

a. quantum mechanics
b. ethnobotany

c. computational biology
d. molecular biology
e. sociobiology

8. Which of the following contains the actual code for building proteins?

   a. phosphate alignment
   b. deoxyribose bonds
   c. base sequences
   d. order of the chromosomes
   e. I-ay ow-knay ut-bay I-ay an't-cay ay-say

9. Which of the following decreases the error rate and simplifies replication?

   a. the radioactivity of DNA ligase
   b. the mutationase enzyme
   c. the mandatory pairing of bases
   d. the double-helix backbone
   e. x-ray crystallography

10. Which of the following individuals is not credited as participating in the discovery of the double helix?

    a. Watson
    b. Mendel
    c. Crick
    d. Franklin
    e. Wilkins

11. Reginald Pisum was exposed to high doses of mutagenic compounds because of a bizarre industrial accident. Should he be concerned about having mutant children?

    a. No; mutations never affect the phenotype.
    b. No; mutations have never been expressed in humans.
    c. Yes; if the mutation occurs in germ-line cells.
    d. Yes; if the mutation affects the autosomal cells.
    e. Yes; if the mutation affects his liver cells, the kids will have mutant livers too.

12. An accident has occurred in the lab. The computer that you have been using to sequence the DNA of yellow-bellied sapsuckers has malfunctioned and you fear that many months of data have been lost. You manage to recover part of a sequence of one strand of DNA (see below) is the gene for this sequence gone for good?

    a. Yes, you need to have both strands in order to know the gene
    b. No, the gene may be on the "surviving" strand
    c. No, if you have one strand, you can determine the sequence of the other
    d. Both b and c
    e. Both a and c

BUT, WHAT'S IT ALL ABOUT?

Here's a question to help you pull together what you've learned so far using this text. If you don't remember how to attack this type of question, check out the example in Chapter 1.

In this chapter, we have seen how a single nucleotide change in the DNA, a point mutation, can be propagated during cell division thanks to the efficiency of DNA replication. Because mutations occur spontaneously during DNA replication, mutations can occur in gamete producing cells as well as the cells of other tissues. Why, then, do heritable mutations that produce a unique phenotype appear only rarely, but mutations that cause cancer, for example, seem more common?

1. **What type of question is this?**
   Compare and contrast? Defend a position? Describe the effect of "A" on "B"?

2. **What kind of evidence do you need?**
   What do you know about DNA replication, mitosis, and gametogenesis? All of these processes factor into this question. What do you know about mutations, genotypes, and effect on phenotype? How can mutations in the genotype produce a phenotype (cancer)?

3. **Pull it all together.**
   Go ahead, exercise those brain cells!

14

CHAPTER 14

How Proteins Are Made: Genetic Transcription, Translation and Regulation

Basic Concepts

- The first step in the process of converting genetic information in protein requires DNA to be transcribed into a form that can carry information out to the protein-synthesis machinery in the cytoplasm. This "messenger" is a molecule of RNA, aptly named messenger RNA (mRNA).
- The second step in the process translates the code held in the mRNA into a sequence of amino acids. The code uses three nucleotide bases to specify a single amino acid, hence the name "triplet code." The code is also redundant; an amino acid may be specified by more than one triplet of nucleotide bases.
- Cells use various signals to regulate protein synthesis; specific sequences within the DNA recognize the signal molecules and determine the rate of protein synthesis.

Transcription

- RNA, ribonucleic acid, is also a polymer composed of ribose sugars holding nitrogen bases and held together by phosphate bonds. It is not a double helix; RNA functions as a single-stranded molecule, and it does not contain thymine, replacing it with the base uracil.
- Complementary base pairing allows an RNA molecule to be synthesized using DNA as a template, allowing the transfer of information from DNA into this mRNA transcript.

THINK ABOUT IT

1. Where does transcription occur, and what is its major product?

2. a. Use the following DNA sequence to build an RNA molecule:

   TACGGTACCATTGCGCAA

b. What process determines which RNA nucleotides match a specific DNA nucleotide?

c. What is this piece of RNA called?

3. What enzyme is responsible for transcription?

Translation

- The mRNA leaves the nucleus and binds to the ribosome, a cytoplasmic organelle where all the molecules needed to make proteins are assembled.
- The mRNA sequence is translated into protein through the action of transfer RNA (tRNA), a small RNA molecule with two functional sites. One site can hold a specific amino acid while the second site "reads" the triplet code of nitrogen bases within the RNA.

THINK ABOUT IT

1. Where does translation occur?

2. What are the roles of mRNA and tRNA in translation?

3. Using the RNA molecule you built from the DNA sequence TACGGTAC-CATTGCGCAA in the preceding "Think about It" section, answer the following questions:

   a. Divide the mRNA into codons. How many do you have?

   b. How many amino acids does the above sequence code for?

4. What is meant by redundancy of the genetic code?

5. Why can the human insulin gene be translated into insulin in bacteria?

6. Draw a eukaryotic mRNA molecule before and after processing.

7. Describe the composition of a ribosome.

8. Draw a sketch of a tRNA that would bring a methionine to the mRNA codon AUG.

9. How does a ribosome "know" that it has reached the end of a given protein?

Genetic Regulation

- DNA contains information, but only when that information is acted upon by proteins can DNA function in the cell. However, controlling whether the DNA may be acted upon is an effective way to control protein translation.
- Jacob and Monod demonstrated that bacteria organize the genes involved in a single metabolic operation into operons; the genes that encode the proteins needed to break down lactose define the *lac* operon. Operons are turned on or off by the binding of small molecules to a sequence of DNA upstream from the start of the operon.

THINK ABOUT IT

1. What does a promoter do?

2. Does every gene necessarily have to have its own promoter? (Hint: Think about the lactose-utilization genes in E. coli.)

3. a. Draw the lac operon.

   b. What is the lac repressor made of?

   c. What would happen to the bacterial cell if the lac operon could not be shut off?

   d. Use the following table to describe the activity of the lactose operon repressor in the presence and absence of lactose.

	Lactose Present	Lactose Absent
Repressor Activity		

4. How many operons do humans have?

5. Do all genes encode proteins?

6. Would a promoter be considered part of a gene?

Cool Fact: Decoding the Triplet Code

To decode Egyptian hieroglyphs, archeologists needed a Rosetta stone—a record that matched hieroglyphs to a known language, Latin. So what was the Rosetta stone that allowed scientists in the early 1960s to interpret the meaning in a sequence of nucleotide bases?

The first "stone" is a mathematical one. Scientists knew there were 20 amino acids, so the DNA code had to be able to call for every amino acid—and it had to be specific, because multiple amino acids could not be specified by a single sequence. Only 16 amino acids could be specified uniquely by a sequence of 2 bases (4×4 possible combinations), so the code must require a longer sequence. A sequence of three bases produces $4 \times 4 \times 4 = 64$ combinations, so the code must involve at least three bases in order to uniquely specify each amino acid. Because a triplet code provides for a great excess of unique sequences, this suggests that the code is redundant—each amino acid may be specified by more than one sequence of three bases.

The second stone takes advantage of the ease of manipulation of bacteria as an mRNA translation system. Bacteria can be gently broken open, and the resulting suspension can be cleared of debris and endogenous DNA and mRNA. Now mRNA with a defined sequence, synthesized chemically, can be added to the bacterial extract along with excess nitrogen bases and ATP for energy. In 1961, Marshall Nirenberg and Heinrich Matthaei added a poly(U) mRNA (a sequence consisting of just uracil bases) to such a bacterial extract and found a polypeptide made of repeated phenylalanine amino acids. They concluded that the sequence UUU must "translate" as phenylalanine. The experiments were repeated with poly(C) and poly(A) messages, determining that CCC codes for lysine and AAA codes for proline.

The researchers determined the rest of the code using a similar procedure, providing painstakingly synthesized mRNA molecules to the cell extract and sequencing the resulting polypeptides. Thus, the Rosetta stone for the genetic code was created using logic and good chemistry.

Chapter 14

TESTING, TESTING

Create a web: polypeptide, gene, protein, enzyme, messenger RNA, DNA, RNA polymerase

Make a flowchart: Follow the stages of protein synthesis.

Vocabulary Review

amino acid DNA intron polypeptide
anticodon exon mRNA rRNA
codon gene operon RNA polymerase tRNA

Transcription	Translation

PRACTICE TEST

After you have finished studying this chapter, close your books, grab a pencil, and spend the next 15 to 20 minutes working on this practice test.

Compare and Contrast

For each of the following paired terms, write a sentence of comparison ("Both . . . ") and a sentence of contrast ("However, . . . ").

transcription/translation

tRNA/mRNA

operator/repressor

intron/exon

Matching

Match the following terms with their description. Each choice may be used once, more than once, or not at all.

a. rRNA
b. lactose
c. anticodon
d. transcript
e. intron

___ Substance that induces the transcription of the lac operon
___ Noncoding segment of DNA
___ Forms part of the ribosome

___ mRNA as it leaves the DNA
___ Part of tRNA

Short Answer

1. Explain the significance of the Jacob-Monod model of the lac operon.

2. How is it that so many different types of proteins can be produced from only 20 amino acids?

3. What is the function of the anticodon?

4. List the differences between DNA and RNA.

Multiple Choice

1. How many mRNA bases make up a codon?

   a. 1
   b. 3
   c. 6
   d. 20
   e. varies, depending on the species

2. What portion of mRNA is removed during processing?

   a. extran
   b. exxon
   c. intron
   d. promoter
   e. codons

3. How do tRNAs bring the correct amino acid to the growing polypeptide?

   a. By base pairing with the rRNA.
   b. By complementary base pairing with the gene in the nucleus.
   c. By covalently binding to the ribosome.
   d. By complementary base pairing between the tRNA anticodon and the mRNA codon.
   e. None of the above.

4. Ribosomes are made up of

   a. protein.
   b. RNA.

c. lipids.
d. a, b, and c
e. a and b

5. Translation of a specific mRNA molecule usually occurs

   a. with multiple ribosomes simultaneously translating.
   b. with many exons loading tRNAs on the codon.
   c. with only one ribosome at a time translating.
   d. with multiple RNA polymerases reading the same gene simultaneously.
   e. with both RNA and DNA nucleotides.

6. What is the correct flow of information from gene to protein?

   a. mRNA-gene-protein
   b. protein-gene-mRNA
   c. gene-protein-tRNA
   d. gene-rRNA-protein
   e. gene-mRNA-protein

7. The end of translation occurs when

   a. an enzyme is sent from the DNA in the nucleus.
   b. endoplasmic substances attach to the ribosome.
   c. a termination codon in the mRNA is reached.
   d. a "telo-tRNA" anticodon attaches to the ribosome.
   e. the translator goes home for the day.

8. Transcription occurs in the _________ while translation occurs in the _________.

   a. nucleus; nucleus
   b. endoplasmic reticulum; nucleus
   c. nucleus; cell membrane
   d. nucleus; cytoplasm
   e. courtroom; classroom

9. An inducible gene is

   a. turned on by DNA nucleotides.
   b. turned on by the presence of its substrate.
   c. turned on at random.
   d. found only in plants.
   e. not found in nature; it is an artificial construct.

10. RNA polymerase

    a. builds transcripts.
    b. excises introns.

c. brings amino acids to the mRNA.
d. forms part of the ribosome.
e. unwinds the double helix.

11. Reginald Pisum is feeling ill, with all the classic symptoms of a head cold. He may even have to stay home instead of attending the Truck & Tractor Pull Semifinals on Saturday night. How is it that one little virus can cause so much misery?

    a. Because viruses are cells, they crowd out the other cells in the body, causing illness.
    b. Viruses convert the protein-synthesis machinery within cells to make many copies of themselves, thereby spreading the infection.
    c. Viruses hate tractor pulls.
    d. Viruses use the lac operon to control Reginald's body cells.
    e. Viruses don't contain thymine, so they interfere with protein synthesis.

12. You go home for Thanksgiving and mention that you have been learning about genetics in biology class. Great Aunt Michele asks you, "What IS a gene anyway?" and you say . . .

    a. "A gene is a molecule that oversees the assembly of amino acids into a polypeptide"
    b. "A gene is a piece of RNA that makes DNA"
    c. " A gene is a segment of DNA that causes the transcription of a segment of RNA"
    d. "A gene is half of a pair of pants made of denim"

BUT, WHAT'S IT ALL ABOUT?

Here's a question to help you pull together what you've learned so far using this text. If you don't remember how to attack this type of question, check out the example in Chapter 1.

Every living thing (and non-living viruses) uses the same triplet code to translate mRNA into proteins. Interestingly, organisms show bias for certain codons; bacteria might prefer to use GCU to code for alanine whereas plants might use GCA. What might account for this codon bias?

What do I do now?

Remember the drill–decide what the question is asking you to do, collect your evidence from this chapter (and the others you've studied) and write!

15

CHAPTER 15

The Future Isn't What It Used to Be: Biotechnology

Basic Concepts

- The tools of biotechnology allow us to produce large amounts of DNA from a small starting sample and to move DNA between organisms.
- Determining the sequence of the human genome will now permit scientists to determine which genes create the proteins that power specific cellular and organismal functions.
- Biotechnology raises many hotly debated, ethical issues.

Tools of Biotechnology

- Restriction enzymes cut DNA at specific, usually asymmetrical sequences. Pieces of DNA cut by the same enzyme can be annealed because they share complementary ends.
- Plasmids and bacteriophages are used as carriers of DNA sequences. DNA from any organism can be cloned into these plasmid or bacteriophage vectors and used to transform a population of bacteria in order to produce the protein product of the cloned DNA.
- Entire organisms can be cloned through the process of reproductional cloning, in which the nucleus from any cell of one organism, such as a sheep, can be introduced into an enucleated egg cell of the same species using a small electrical current to simulate fertilization. The resulting embryo will be a genetic copy of the organism that donated the DNA (nucleus).
- Copies of DNA sequences can be made using a small amount of DNA to serve as the template in the polymerase chain reaction (PCR). PCR increases the amount of DNA in a sample so that other analyses can be performed, and it has proved to be an invaluable technique for criminal investigations, disease diagnosis, and evolutionary studies.

THINK ABOUT IT

1. What is the natural source of HGH? In what organism is HGH produced by biotechnology?

2. How can a human protein be made by a bacterium?

3. What do restriction enzymes do?

4. What type of molecules serve as "suitcases" for carrying genes into bacteria?

5. Outline the five steps required to get a human gene to make protein in bacteria.

6. What is cloning? How would you clone your best friend or dog?

7. Do cloned organisms arise by mitosis or meiosis?

8. What do you think the impact of PCR has been on the field of forensics?

9. Does PCR make copies of genes or proteins?

10. You cut a piece of DNA into three fragments: 500 base pairs, 1000 base pairs, and 2000 base pairs in length. Mark the following diagram of an agarose gel to indicate the relative positions of these three fragments.

negative electrode

positive electrode

11. Why would you want to know the sequence of a piece of DNA?

The Human Genome Project

- Using cloning, PCR, and electrophoretic and sequencing techniques, the nucleotide sequence of 90 percent of the roughly 3-billion-base-pair human genome was published in June 2000—the result of a collaborative effort of a private biotech company and a publicly funded consortium of 16 laboratories.
- The result of the Human Genome Project has been compared to a phone book. Scientists now know the "names" and "addresses" of every gene within the human genome, but not what the genes do.

- Matching genes to their functions is part of the growing field of genomics; matching genes to their protein products, and protein products to function, is the domain of proteomics.

THINK ABOUT IT

1. Does knowledge about the sequence of every gene in the genome tell you what every gene does in the human body?

2. How could the human genome sequence help a medical geneticist studying an inherited genetic disorder?

Biotechnology and Ethics

- Recombinant DNA cloning makes it possible to produce large quantities of pharmaceutically important human biomolecules, such as insulin or growth hormone. The availability of some hormones could lead to abuse.
- Cloning facilitates the production of transgenic plants, crops altered to increase disease and pest resistance or to yield greater nutritive value. The effect of these plants upon wild versions or beneficial insect species is uncertain.
- The potential use of biotechnology to modify our own species, either through reproductional cloning or recombinant DNA cloning, is an unresolved issue. Just because we can modify organisms, does that mean we should modify them?

THINK ABOUT IT

1. What would be the impact of gene therapy for a hemophilia patient who currently receives injections of a blood-clotting factor?

2. How will PCR be used to decipher the Dead Sea Scrolls?

3. Which of the following would you consider to be acceptable uses of biotechnology?

   a. providing everyone with the opportunity to have blue eyes

   b. providing doctors with the opportunity to correct, in babies, certain eye defects associated with brown eyes

   c. providing crops with frost resistance to save farmers billions in lost revenues each year

   d. releasing a frost-resistance gene into the environment that allows roses to bloom year round

e. allowing 60-year-old women to bear children

f. allowing 30-year-old women (with 60-year-old husbands) to bear children

4. What is genetic discrimination?

Cool Fact: Disease and Genetics

Diseases caused by a defect in a single gene are extremely rare. However, most scientists believe that even common diseases such as cancer, heart disease, alcoholism, and schizophrenia have a genetic component. Genetics would explain why one lifelong smoker develops cancer but another doesn't. But which genes are involved? Scientists could answer this question if they could show that certain conditions seem to "run in families" and that the family members who get the disease all have the same alleles for the genes of interest. Thus, "proof" of a genetic component for a common disease requires many detailed family trees—for tracing "inheritance" of the disease state—and a genetic map for each family member.

The Human Genome Project may provide such proof. The people of Iceland have kept detailed genealogical records for centuries. Iceland has a reasonably isolated population; because few "outsiders" marry into the population, its genetic diversity is rather limited. The country also has a well-organized system of socialized medicine. More than 80 percent of the population of Iceland has agreed to allow de Code Genetics, a for-profit company, access to their medical and genealogical records to create a "map" of disease inheritance. Using this map and the map of the human genome, de Code Genetics plans to search for genes that occur not just in a single family, but in anyone with a specific disease. The power of de Code Genetics' approach is not only being able to trace the presence of a gene in previous cases of disease but also having the ability to predict the occurrence of disease in future generations.

This approach is not without controversy. Even though de Code Genetics has garnered public permission to use medical records, many people believe it is inappropriate to use "private" property and public knowledge to make money. Moreover, what will be the consequences of people learning that they are "fated" to develop heart disease or schizophrenia? Is it enough to have all of the "wrong" genes, or is an environmental trigger necessary too? These are questions that must be, and will be, hotly debated in the near future.

Chapter 15

TESTING, TESTING

Create a web: DNA, restriction enzymes, tandem repeats, forensic DNA typing, blood, RNA, gel, size

Make a flowchart: Describe the process for making a bacterial cell produce the human hairy-ears protein.

Vocabulary Review

bacteriophage

biopharmaceutical

forensics

gene therapy

Human Genome Project

plasmid

recombinant insulin

restriction enzymes

vector

Cloning Tools	Applications

PRACTICE TEST

After you have finished studying this chapter, close your books, grab a pencil, and spend the next 15 to 20 minutes working on this practice test.

Compare and Contrast

For each of the following paired terms, write a sentence of comparison ("Both . . . ") and a sentence of contrast ("However, . . . ").

gene cloning/reproductive cloning

cloning vector/plasmid

polymerase chain reaction/DNA sequencing

proteomics/genomics

Matching

Match the following terms with their description. Each choice may be used once, more than once, or not at all.

a. restriction enzyme
b. recombinant DNA
c. PCR
d. vector
e. gel electrophoresis

___ A molecule used to carry foreign genes into bacteria
___ A rapid way to amplify DNA in the laboratory
___ A way to separate DNA fragments based on their size

___ Molecular scissors
___ A gene sequence from more than one origin

Short Answer

1. Explain xenotransplantation.

2. Electrophoresis gels are used to separate molecules by two criteria. What are they?

3. How can forensic DNA typing (aka "DNA fingerprinting") determine the guilt or innocence of a suspect?

4. Why is a percentage of funding for the Human Genome Project devoted to bioethics?

Multiple Choice

1. Which molecules are analogous to DNA scissors?
   a. plasmids
   b. vectors
   c. restriction enzymes
   d. human growth hormone
   e. none of the above

2. What is the goal of the Human Genome Project?
   a. to generate restriction fragments for the entire human genome
   b. to cure all human genetic disease
   c. to sequence every person's genome and enter it into a database
   d. to determine the base sequence of the entire human genome
   e. to develop new genes to introduce into humans

3. Dolly the sheep is
   a. an exact copy of her father
   b. an animal with DNA from two different cells in each cell of her body
   c. a reproductive clone
   d. haploid
   e. c and d

4. What is gene therapy?
   a. a way to introduce good copies of genes into a person's body
   b. the injection of an active protein into a patient with a genetic disorder
   c. a science-fiction fantasy
   d. the consumption of genes as an energy source
   e. the transformation of bacteria with a recombinant plasmid

5. You have three copies of a particular DNA molecule. What technique would you use to make more copies of the molecule?

   a. gel electrophoresis
   b. sequencing
   c. PCR
   d. restriction fragment analysis
   e. none of the above

6. Having become an expert on gel electrophoresis, you are asked to examine a gel for a colleague. Where would you find the smallest segments of DNA?

   a. near the positive electrode, farthest away from the wells
   b. near the negative electrode, close to the wells
   c. near the top, near the negative pole
   d. near the middle; they tend to slow down after the first few minutes
   e. in the wells

7. After donating a small amount of blood to a research study, Reginald Pisum has received a rather disturbing phone call. It seems the study has shown that humans (as represented by Reginald) and Mendel's pea plants (Pisum sativum) share a large number of genes. This study was in what area of molecular biology?

   a. tectonics
   b. proteomics
   c. economics
   d. genomics
   e. platonics

8. Vectors

   a. deliver DNA.
   b. cut DNA at specific sites.
   c. paste DNA fragments together.
   d. read DNA and display its sequence of nucleotides.
   e. make DNA from RNA.

9. Recombinant DNA

   a. requires the use of PCR to make.
   b. is harvested from viruses.
   c. contains no thymine bases.
   d. demonstrates the universality of the genetic code.
   e. naturally occurs in most plants.

10. Biotechnology

    a. is an ancient science.
    b. has been producing mammal clones for about 50 years.
    c. can cure every illness.
    d. produces only drugs.
    e. is so new that regulation has not caught up with it yet.

11. Bacterial plasmids are of particular interest to biotechnologists because

    a. plasmids cause random mutations within human and other eukaryotic cells.
    b. plasmids glow spontaneously when used in gel electrophoresis.
    c. plasmids make the DNA of their cells rigid like plastic and easy to identify.
    d. plasmids can deliver DNA to other cells by the process of transformation.

12. Back to the household of Mr. and Mrs. Pisum. The Pisums decide to create a child by cloning, using their neighbor, Mrs. Droso Melanogaster as an egg donor. According to the procedure outlined in your text, the resulting offspring will be an exact genetic copy of

    a. Mrs. Melanogaster, who donated the egg.
    b. Mrs. Pisum, who donated the cell with DNA.
    c. Mr. Pisum, who donated the sperm.
    d. Mrs. Peacock, in the library with the rope.

BUT, WHAT'S IT ALL ABOUT?

Here's a question to help you pull together what you've learned so far using this text. If you don't remember how to attack this type of question, check out the example in Chapter 1.

Imagine that it is September, 2001, and that you are employed by the Coroner's Office in New York City as a forensic technician. You are surveying the site of one of the most devastating mass disasters our country has ever witnessed, and your job is to officially identify the victims so that insurance claims and other legal processes can be settled. What are you going to do to certify death certificates for thousands of people whose remains are fragmentary?

What do I do now?
Remember the drill – decide what the question is asking you to do, collect your evidence from this chapter (and the others you've studied) and write!

16

CHAPTER 16

An Introduction to Evolution: Charles Darwin, Evolutionary Thought, and the Evidence for Evolution

Basic Concepts

- All living things on Earth have descended from a common ancestor; modifications of ancestral forms over extraordinarily long time periods have produced the great variety of living species now in existence.
- Living things, humans included, are neither static nor directed on a particular course of development. Natural selection drives evolution; that is, the interaction of the living organism with the nonliving environment drives change blindly.
- Five independent lines of evidence support Darwin's theory of descent with modification, the cornerstone of evolutionary theory.

Common Descent with Modification

- Living organisms possess a suite of traits and behaviors that determine their survival and reproductive success. Collections of traits that enhance reproductive success allow an organism to leave more surviving offspring, who pass those traits on to future generations.
- The pressure of natural selection over long periods of time causes the descendant organisms to look different from the ancestral parent, because different traits have proven more useful to survival and reproductive success.
- Although Charles Darwin was not the only person to describe the process of descent with modification, he provided observational evidence supporting the hypothesis of common descent with modification and first described the selection pressure exerted by natural forces.
- Darwin's thinking about the origin of species was influenced by Charles Lyell's geological observations suggesting that the Earth itself was not static, by Georges Cuvier's fossil evidence, and by T. R. Malthus' writings about human populations.

THINK ABOUT IT

1. What is the principle of common descent with modification?

2. What is meant by natural selection?

3. From how many living species did all living things on Earth descend?

4. a. What was Darwin's job aboard the HMS *Beagle?*

   b. Was that what he had originally trained for at Cambridge?

5. Given the following timeline, where would you place these events?

   A - Darwin's birth

   B - Voyages on the HMS *Beagle*

   C - Publication of *On the Origin of Species*

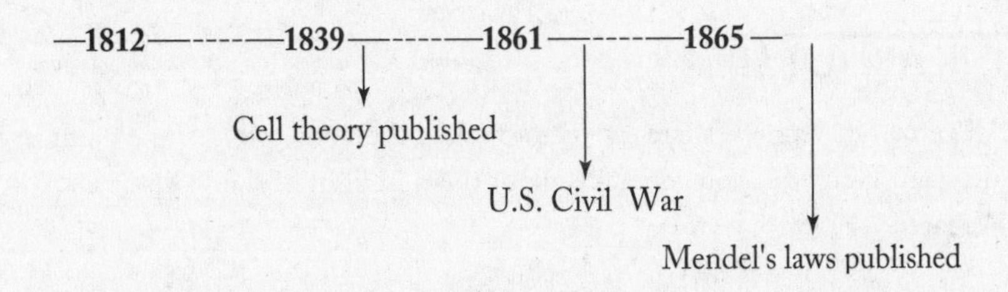

6. What did Lamarck propose as an evolutionary mechanism?

7. What was Cuvier's observation regarding the fossilized remains of species?

8. How did Darwin perceive the formation of new species through natural selection?

9. What was surprising about the birds Darwin brought back from the Galapagos Islands?

10. Malthus provided a final puzzle piece in the construction of Darwin's theory; what was it?

11. What did Wallace view as the force behind evolution?

12. What would you do if you were placed in Darwin's situation and asked to comment on another's work nearly identical to your own?

Multiple Lines of Evidence for the Theory of Evolution

- Radiometric dating, which uses the "clock" of radioactive decay to measure time, indicates the Earth is about 4.6 billion years old—old enough to provide the time for evolution.
- Fossil evidence shows the change from "simple" organisms in old layers of rock to more complex organisms in more recent layers, as the theory of evolution would predict.
- Embryology (the study of animal development) and morphology show that different-looking structures in adult animals develop from common structures in the embryo, as we would expect if animals had developed from a common ancestor.
- All organisms share a common genetic code, which is strong evidence for a common ancestor.
- The predictions of evolutionary theory are borne out in laboratory experiments using populations of rapidly reproducing organisms, like fish and bacteria.

THINK ABOUT IT

1. All vertebrate animals share a structure called the notchord at some point in their life span. What can you conclude about them in light of the work of Darwin and Wallace?

2. How did acceptance of evolution explain embryonic gill slits in terrestrial species such as humans and chickens?

3. Was evolution or natural selection less accepted?

4. What is actually passed on to subsequent generations in the process of evolution?

5. How does the Modern Synthesis differ from Darwin's theory of evolution?

6. What groups of scientists contributed to the Modern Synthesis?

7. When (approximately) did the work of Darwin and Wallace become generally accepted among scientists?

8. How do we know the approximate age of fossils found in particular layers of rock?

9. What is a homologous feature? What evidence do homologous features provide in support of evolution of species from a common ancestor?

10. You look at the sequence of rRNA molecules from three different organisms (A, B, and C). A and B differ at 5 bases, and A and C differ at 3 bases. Is B or C more closely related to A?

11. In the following space, give the four lines of evidence that support Darwin and the Modern Synthesis. Be sure to list the type of evidence used by each line.

Evidence:	**Support:**

12. Several species of mammals have been grouped by scientists as the Canids (wolf, coyote, fox, wild dogs, hyena, domesticated dogs, and so on).

    a. Explain how the core principles of evolution, descent with modification and natural selection, apply to the Canids.

    b. What four kinds of evidence do you think the scientists used to create this grouping?

Cool Fact: Scientific Communication

So why do we give credit for the theory of evolution to Darwin, and not Wallace? Wallace wrote up his theory and sent it to Darwin for comment; doesn't that mean that Darwin "stole" Wallace's idea? Assigning credit for significant accomplishments in science lies at the heart of the scientific communication.

Credit for an idea falls to the first person(s) whose documentation, meaning observations or experimental evidence, of the idea appears "in print." Electronic publishing has changed the face of scientific communication because research articles may appear online long before the hard copy hits the library shelves. Regardless of whether a document is published in electronic or paper form, the work must undergo a rigorous evaluation by other scientists before it is deemed "worthy" of publication. Outside reviewers, scientists in the field and familiar with the work being done, read the paper and evaluate its merit. Does this study represent new knowledge? Is the study done appropriately; have the authors adequately described their methods and used the proper controls? Can the authors justify their conclusions based on the data they've accumulated, or are more experiments required? Before being accepted for publication, a study must first pass the scrutiny of its reviewers, and the author must address reviewers' comments.

Wallace's paper describing his theory of natural selection was presented, along with a short paper by Darwin, at a meeting of the Linnean Society. But Darwin then published his *On the Origin of the Species* before Wallace returned from Indonesia. One could argue that because Darwin had not published his idea during the twenty years after his voyage, and was forced to publish only after Wallace wrote his paper, that Wallace truly deserves primacy on the theory of natural selection. Perhaps he does, but as a man with a working-class background in nineteenth-century England, Wallace lacked the credentials to pursue his claim against Darwin, an educated man of the upper middle class with a history of publications on natural science. Sometimes factors other than good science affect how science history is recorded.

My thanks to Dr. William Coffman, my colleague at the University of Pittsburgh, who provided me with the details of the Wallace-Darwin story presented here.

Chapter 16

TESTING, TESTING

Make a web: Darwin, Wallace, natural selection, evolution, individual, species, common descent

Make a flowchart: Chart the contributors to the Modern Synthesis of Evolution.

Vocabulary Review

common descent with modification | fossil record | limited resources
comparative embryology | homologous features | natural selection

Evolutionary Mechanisms	Evidence for Evolution

PRACTICE TEST

After you have finished studying this chapter, close your books, grab a pencil, and spend the next 15 to 20 minutes completing this practice test.

Compare/Contrast

For each of the following paired terms, write a sentence of comparison ("Both . . .") and a sentence of contrast ("However, . . .").

paleontology/taxonomy

natural selection/fitness

embryology/homologous structures

evolution/natural selection

Matching

Match the following terms with their description. Each choice may be used once, more than once, or not at all.

a. proposed that species evolved by the inheritance of characters acquired by their parents
b. proposed that geological processes are still ongoing
c. provided the icing on the cake for Darwin's theory
d. wrote letter proposing natural selection and common descent with modification as the mechanism for evolution
e. sailed on the HMS *Beagle*

___ Darwin
___ Lamarck
___ Lyell
___ Malthus
___ Wallace

Short Answer

1. Explain descent with modification.

2. What English naturalist is named as the co-discoverer of Darwin's theory?

3. Has evolution ever been successfully demonstrated in a scientific experiment?

4. Where did Darwin gather the evidence that he used to define his theory of evolution?

Multiple Choice

1. We would predict that two very distantly related species like bluebirds and alligators probably

   a. share most of the same DNA sequences.
   b. share more DNA sequences than a bluebird and a duck.
   c. share relatively few DNA sequences.
   d. share all the same DNA sequences.
   e. sharing is optional, giving is wise.

2. Is the layering of fossils in the Earth used to support evolution?

   a. Yes; fossils are laid down in chronological order and therefore we can observe changes in these species over time.
   b. No; the fossil evidence is too recent to provide support.
   c. No; because Darwin didn't use it, the Modern Synthesis also ignores it.
   d. Yes; the fossil evidence is the only support for Darwin's idea of acquired characteristics.
   e. Yes; the fossil record provides evidence for how present-day (extant) species have changed their embryonic developmental patterns in the past 100 years.

3. You take a job as a research assistant at one of the richest fossil fields ever to be discovered. On your first day, you discover a new species, *Neko borgus,* that bears similarities to the modern domesticated cat. The bed where this fossil was found also contained several fossils of early primates. From this you conclude that

   a. *Neko borgus* must have been active millions of years before the primates appeared.
   b. *Neko borgus* must have been active millions of years after the primates appeared.
   c. *Neko borgus* must have been active at about the same time as the primates appeared.
   d. The mixing of fossils was likely due to geological events like those described by Charles Lyell.
   e. Early primates kept domesticated cats.

4. What is the supportive evidence for evolution from comparative anatomy?

   a. All antelopes have four legs.
   b. Most animals have a head.
   c. Closely related species have similar structures.
   d. Similar gene sequences code for similar proteins.
   e. Diverse species have homologous structures.

5. What is the supportive evidence for evolution from comparative embryology?

   a. All plant seeds look alike.
   b. All embryos arise from the union of egg and sperm.
   c. Different species have different embryos.
   d. Different species develop along the pattern set by their common ancestor.
   e. All baby animals look alike.

6. Which of the following provide evidence for the process of evolution?

   a. radiometric dating
   b. the fossil record
   c. molecular biology
   d. experiments on guppies
   e. All of the above.

7. What is a consequence of natural selection?

   a. Species become extinct.
   b. Individuals that fit the environment better leave more offspring.
   c. Certain species will be preserved in the fossil record.
   d. All organisms share a common ancestor.
   e. b and d

8. How many life forms must have existed when life originated on Earth?

   a. as many as every living species today
   b. as many as every species that is living and has ever lived on Earth
   c. one
   d. three—a bacterial ancestor, a plant ancestor, and an animal ancestor
   e. approximately the population of the Earth today

9. Which of the following scientists is not correctly matched with his theory?

   a. Cuvier—continuous fossil record with no apparent breaks
   b. Darwin—natural selection as a mechanism for evolution
   c. Wallace—natural selection as a mechanism for evolution
   d. Lamarck—characters acquired during an organism's lifetime are passed to offspring
   e. Lyell—dynamic geological events continue to shape the Earth

10. Which of the following are examples of evolution?

    a. different finch species living on different Galapagos islands
    b. the rise of antibiotic-resistant strains of bacteria
    c. changes in guppy populations after the introduction of predators
    d. All of the above.
    e. None of the above.

11. How does the Modern Synthesis differ from Darwin's Theory of Evolution?

    a. The Modern Synthesis includes the evidence from several lines of scientific research.
    b. The Modern Synthesis was first proposed by Alfred Russell Wallace.
    c. The Modern Synthesis was introduced by Watson and Crick after the discovery of DNA structure.
    d. The Modern Synthesis was the term the Darwin first used to describe descent with modification.

12. Which of these would NOT be homologous to the others?

    a. arm bones of a human
    b. front leg bones of a dog
    c. front fins of a fish
    d. bones of a dolphin flipper

BUT, WHAT'S IT ALL ABOUT?

Here's a question to help you pull together what you've learned so far using this text. If you don't remember how to attack this type of question, check out the example in Chapter 1.

Imagine the Earth in one million years. We're gone, but our legacy of global warning remains. Regions of the world that were semi-tropical in the twenty-first century are now tropical, and artic regions are more temperate. Given what you know about genetics, natural selection, and common descent with modification, what is your prediction for life on Earth under these conditions?

What do I do now?

Remember the drill—decide what the question is asking you to do, collect your evidence from this chapter (and the others you've studied) and write!

CHAPTER 17

The Means of Evolution: Microevolution

Basic Concepts

- Populations evolve because of a change in the frequency of alleles within the gene pool.
- The five agents of microevolution are (1) mutation, (2) migration, (3) genetic drift, (4) non-random mating, and (5) natural selection.
- Natural selection acts on individuals within a population, determining the direction of evolution for the population as a unit.

Evolution and Populations

- Evolution, which is the change in form and function of a population over many generations, is driven by changes in DNA.
- Because a diploid organism has two alleles for any given gene, many allelic variations of a gene can coexist within a population. These allelic variations translate into small phenotypic variations among members of the population.
- Environmental pressures select an optimum phenotype for survival, thus increasing the proportion of alleles in succeeding generations that encode for a successful phenotype. Over many generations of natural selection, the frequency of a given allele within a population's gene pool may increase or decrease.

THINK ABOUT IT

1. What is the smallest evolving unit?

2. What does evolution act on?

3. Which of these is the largest—species, population, or individual?

4. What is microevolution?

5. What is the driving force behind microevolution?

6. a. A moth population in England gradually acquired darker coloration during the Industrial Revolution. Is this an example of macro- or microevolution?

   b. Whales are thought to have evolved from land-dwelling organisms. Is this an example of macro- or microevolution?

   c. Many human sports use a round object (ball); is this macro- or microevolution?

   d. U.S. and Canadian football follow different rules; is this macro- or microevolution?

Five Agents of Microevolution

- Mutations, because they change the information encoded in DNA, can drive evolution. However, mutations that add new, adaptive information are extremely rare, so mutation is a minor agent of evolutionary change.
- More commonly, alleles are added to or removed from a population by migration of members into or out of the population. This gene flow can be a major driving force for change if the new members represent the contribution of a different gene pool.
- Small populations may be significantly affected by the random loss of alleles due to disease, natural disaster, or migration into a new habitat. In these cases, alleles are not selected by environmental factors; they are lost without regard to their adaptive value. Such losses cause genetic drift within a population—"drifting" toward a change in allelic frequencies.
- Allelic representation in future generations can change if ancestral organisms carrying that gene mate more frequently than those that don't carry the gene. Non-random mating occurs during sexual selection by females for a particular male phenotype; males possessing the desirable phenotype mate more, thus increasing the frequency of the alleles for that phenotype in subsequent generations.
- Natural selection forces populations to adapt to changing environmental conditions by favoring the survival of the organisms within a given population that best "fit" in the environment at a given time. "Fitness," then, is a relative concept; it can be defined for an individual only relative to other members of the population, and only for a given set of environmental conditions.

THINK ABOUT IT

1. Complete the following table describing the factors influencing genotype frequencies in a population, how they exert their effect, and whether the effect is adaptive or not (random).

Factor:	How effect is exerted:	Is effect adaptive?
Mutation		
Gene flow		
Genetic drift		
Non-random mating		
Natural selection		

2. Is genetic drift likely to have a greater effect in a small or large population? Explain your answer.

3. Populations decrease in size because of two different effects. Indicate what each effect is called by biologists, and give an example of each type of effect. One effect happens to populations that are new to a place, and the other one happens to existing populations.

   a. Effect: ________________ Example:

   b. Effect: ________________ Example:

4. Given the following list of molecules, where would a genetic mutation first appear? In which molecule would that mutation appear next?

   DNA polypeptide tRNA mRNA rRNA

5. Give an example of how gene flow has affected the human population of North America both genetically and culturally.

6. Label the following statements with an F (founder effect) or B (bottleneck effect):

   a. ____ Decreased genetic variability due to an epidemic that wipes out 84 percent of the population.

   b. ____ Decreased genetic variability because the pioneers of the new population did not represent the diversity of the original population.

7. How would you describe fitness in an evolutionary sense?

8. Can the characters responsible for fitness in a population be absolutely defined?

9. Who is more fit, an eagle or a field mouse? Why?

10. How can you tell which individuals in a population have the greatest relative fitness?

Three Modes of Natural Selection

- Stabilizing selection acts on polygenic characteristics by favoring "average" forms over the extremes, such as in the case of infant birth weights.
- Directional selection acts to move a population toward expressing one of the extreme forms of a phenotype, as in the case of the moth that became darker in response to a sooty environment.
- Disruptive selection drives a character toward both of its extreme forms, a rare occurrence in nature.

THINK ABOUT IT

1. Caterpillars that can maintain coloration similar to leaves avoid predation by birds. What kind of selection is this?

2. You have $1.00 in change in your pocket. What would be the composition of coins for each of the following "coin selection" regimes?

   a. stabilizing selection

   b. directional selection

   c. disruptive selection

3. Label the following statements according to the type of selection at work. (**S**—stabilizing, **D**—directional, **R**—disruptive)

   ____ The smallest birds in the flock are most likely to be eaten by predators.
   ____ Shoe manufacturers stop making shoes for very small and very large feet.

____ Space aliens abduct only those too big to hide and too small to fight.
____ Over the past few centuries, average heights of human females in North America have increased.

Cool Fact: Microevolution on a Human Scale

Where does disease come from? Why do pathogens make us sick but cause few problems in their animal hosts? How do pathogens make the switch from animal to human hosts? Does human evolution drive parasite evolution, or is it the other way around? Answers to such questions may help scientists and physicians fight seemingly endemic diseases, such as malaria and AIDS. Those answers may come as scientists use DNA sequence data to map the evolutionary trees of pathogenic parasites, bacteria, and viruses.

Fossils of single-cell organisms don't exist, so building a family tree relies upon comparing DNA sequences of closely related organisms that infect different hosts or exist in geographically distinct regions of the world. Using this technique, scientists traced the evolution of trypanosomes, the parasite responsible for sleeping sickness. By comparing the sequences of many Trypanosoma species, scientists found that the trypanosome causing sleeping sickness (endemic to Africa) and the trypanosome responsible for Chagas' disease (endemic to South America) shared a common ancestor during the time of the supercontinent, Gondwanda.

Some heritages are not so ancient, but still older than most scientists believed. One model suggests that many diseases made the jump from animal hosts to human hosts when humans began living in "civilized" societies 10,000 years ago. Many people living in proximity provide an easy journey for a parasite from host to host. Surprisingly, some diseases may have evolved before human hosts lived in societies. Bacteria in the genus Shigella cause dysentery, a common disease in the developing world that can produce a lethal, bloody diarrhea. Because the dysentery-causing bacteria inhabit only humans, scientists believed that this pathogen arose as human societies formed. Genetic evidence, however, suggests that Shigella is closely related to E. coli, a relatively harmless bacteria living in the human intestine. These bacteria are so closely related that Shigella may not even be a separate genus, but rather several strains of E. coli that have accumulated the genes that make invading the intestinal cells possible. Shigella strains appear to have differentiated from E. coli between 27,000 to 35,000 years ago, long before the first recorded human societies.

That pathogens adapt to their human hosts is not surprising; but the fact that some of them did so long before humans represented an "easy" host is remarkable.

Chapter 17

TESTING, TESTING

Make a web: genes, alleles, gene pool, population, microevolution, fitness, reproductive success

Make a flowchart: Show how each of the five agents of microevolution listed in the text can have an effect on allelic frequencies within the gene pool.

Vocabulary Review

Place the following terms in the appropriate columns of the table. Some terms may be used in more than one column.

allele	founder effect	genetic drift	natural selection
bottleneck effect	gene flow	genotype	non-random mating
fitness	gene pool	migration	population

Agents of Microevolution	Targets of Natural Selection

PRACTICE TEST

After you have finished studying this chapter, close your books, grab a pencil, and spend the next 15 to 20 minutes completing this practice test.

Compare and Contrast

For each of the following pairs of terms, write one sentence describing their similarities and one sentence explaining their major differences. In writing about similarities, start the sentence with "Both . . . "; for differences, start with "However . . . "

microevolution/macroevolution

gene flow/genetic drift

gene pool/population

fitness/adaptation

Matching

Match the following terms with their description. Each choice may be used once, more than once, or not at all.

a. genetic drift
b. directional selection
c. gene pool

d. microevolution

e. fitness

____ Favors the extreme range of a phenotype

____ Change in gene frequency in a small population

____ The total alleles existing in a population

____ Relative measure of the ability to reproduce

____ Genetic change in populations causing new phenotypes to appear

Short Answer

1. Both mutation and natural selection can change the allelic frequencies within the population gene pool. Explain the difference in their effects.

2. What is the difference between the bottleneck effect and the founder effect?

3. How can you determine which of two individuals is more "fit" in the evolutionary sense?

4. How can sexual selection affect the frequency of alleles within the gene pool?

Multiple Choice

Circle the letter that best answers the question.

1. Which of the following can contribute to genetic drift?
   a. small population size
   b. a population bottleneck
   c. the founder effect
   d. a and b only
   e. All of the above.

2. Assortative mating is based on
   a. allele frequency.
   b. phenotype.
   c. the founder effect.
   d. randomness of mate selection.
   e. None of the above.

3. Back in the fossil field, you have discovered many fossils that are similar to *Neko borgus*. You notice however, that the more recent Neko fossils tend to be larger than those that are more ancient. From this you conclude that
   a. Neko has been subjected to directional selection.
   b. Neko species continue to grow after death.
   c. Neko has been subjected to stabilizing selection.
   d. Environmental conditions have caused the Neko fossils to expand over time.
   e. Neko has been subjected to disruptive selection.

4. Evolution is evidenced in the

   a. genes of an individual.
   b. phenotypes of individual organisms.
   c. allele frequencies in a population.
   d. birds with small beaks.
   e. birds with large beaks.

5. What is the consequence of genetic drift?

   a. improvement of the population
   b. increase in strength of the population
   c. decrease in IQ of the population
   d. unpredictable effects on characters in a population
   e. increase in IQ of the population

6. Which of the following processes is/are adaptive?

   a. genetic drift
   b. mutation
   c. gene flow
   d. natural selection
   e. All of the above.

7. A shuffling of alleles like a deck of cards leads to different "hands," known as

   a. genotypes.
   b. genes.
   c. flushes.
   d. fossils.
   e. flavors.

8. The emergence of very diverse species of house pets (birds, domestic dogs) is an example of

   a. primary evolution.
   b. secondary evolution.
   c. macroevolution.
   d. microevolution.
   e. minievolution.

9. Of the following, which consistently pushes populations toward a better "fit" with their environment?

   a. nonrandom mating
   b. genetic drift
   c. natural selection
   d. gene flow
   e. founder effect

10. The relative fitness of different types within a population is probably

    a. unchanging.
    b. permanent.
    c. changeable.
    d. both a and c
    e. All of the above.

11. The number of individuals within a population with a certain genotype is predicted by

    a. Mendel's Law of Segregation.
    b. Darwin's Scale of Biodiversity.
    c. the Hardy-Weinberg Theorem.
    d. the number of angels who can dance on the head of a pin.

12. What prevents a population from becoming genetically isolated?

    a. genetic drift
    b. gene flow
    c. bottleneck event
    d. any of these could prevent genetic isolation.

BUT, WHAT'S IT ALL ABOUT?

Here's a question to help you pull together what you've learned so far using this text. If you don't remember how to attack this type of question, check out the example in Chapter 1.

Scientists have puzzled about the evolutionary processes needed to generate complex structures, like eyes for example. Could a rudimentary eye (a patch of cells that detect light) have developed slowly by accumulating mutations, or suddenly, by many mutations occurring all at once? Given what you know about mutations and natural selection, do you think complex structures develop slowly or suddenly?

What do I do now?

Remember the drill—decide what the question is asking you to do, collect your evidence from this chapter (and the others you've studied) and write!

18

CHAPTER 18

The Outcome of Evolution: Macroevolution

Basic Concepts

- The basic unit of life forms is the species—a natural population of interbreeding organisms that are reproductively isolated from other populations of organism that they could or do interbreed with.
- New species occur when populations cease to interbreed through the development of some reproductively isolating mechanism.
- Each species can be identified by a unique, binomial name and placed in a taxonomy according to the degree of relatedness to other species.

Biological Species Concept

- Species are groups of natural populations that are reproductively isolated from other species. Thus, breeding behavior is the defining feature of a species.
- New species arise when populations of interbreeding organisms become reproductional after being isolated from each other. The parental species may be transformed over time into a dramatically different species through the action of natural selection (anagenesis), or a new species may branch off from the parental species due to some force that reproductionally isolates the offspring from the parents (cladogenesis).

THINK ABOUT IT

1. What does the biological species concept use to define a species?

2. Under what situations, or for what organisms, does the biological species concept not work?

3. The sequences of the rRNA genes from bacterial isolates A, B, and C were determined. A and B differ at 20 nucleotides; A and C differ at 45 nucleotides; and B

and C differ at 5 nucleotides. None of these isolates mates with one another. Are any of these isolates likely to be the same bacterial species?

4. Which diagram (top or bottom) represents speciation by cladogenesis, and which represents speciation by anagenesis?

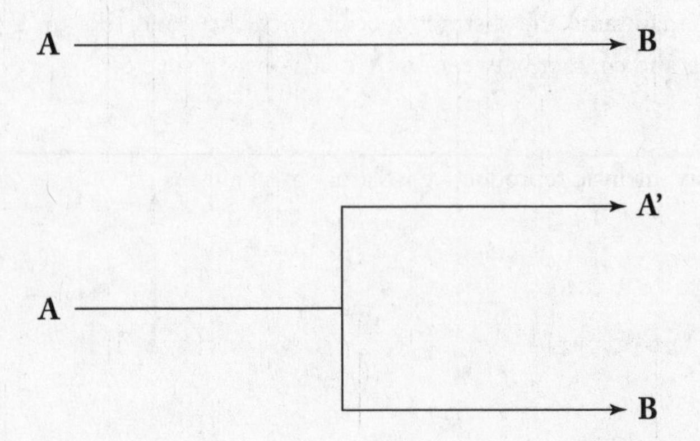

5. What must happen to populations for cladogenesis to cause branching into two species?

Reproductive Isolating Mechanisms

- Geographic separation is necessary to achieve speciation, but it must be accompanied by some other isolating mechanism to assure that the separated populations accumulate enough phenotypic changes while separated to make interbreeding impossible should the two populations be reunited. Geographic separation is an extrinsic isolating mechanism.
- Six intrinsic reproductive isolating mechanisms foster the process of speciation in the absence of geographic isolation. Closely related organisms that occupy different habitats, that do not mate in the same season, or that share courtship rituals will remain separate species despite geographic proximity. Physical differences between the two related species can encourage speciation, such as disparities in genital organs that prevent mating, genetic incompatibilities between gametes, or infertility of the resulting zygote.
- Speciation is more likely to occur in a species that has a specialized lifestyle in a given environment. Likewise, environments that provide many resources and habitats to exploit encourage speciation.
- The fossil record suggests that speciation occurs rapidly, with long periods of stability between speciation events.

THINK ABOUT IT

1. What kind of separation occurs in allopatric speciation?

2. Suppose that two populations of pond plankton are separated from one another by the drying of their pond into two large puddles and then go on to speciate into different species.

   a. What is the mechanism of speciation?

   b. What maintains the distinct species once the pond fills up again, allowing mixing and contact between individuals of each species?

3. List the six intrinsic reproductive isolating mechanisms.

(a)	(d)
(b)	(e)
(c)	(f)

4. Describe sympatric speciation.

5. Write the term that describes each of these situations.

   a. two populations; one feeds on fruit ripening in July, the other on fruit ripening in September

   b. two populations that have a great difference in genital size

   c. individuals from two populations that interbreed but whose offspring are always sterile

The Categorization of Earth's Living Things

- Taxonomic classification groups species into categories depending on the degree of relatedness. Related species share a common ancestor, and the degree of relatedness depends on how long ago that ancestor lived.
- Classification requires that each species have a specific and universally understood name. The binomial (two-name) naming system was developed by Carl von Linné (Linnaeus, in Latin) in the eighteenth century.
- Classical taxonomy uses physical structure (morphology) to fit animals within the evolutionary tree, comparing modern forms with fossil forms to determine relatedness. DNA sequence information also provides an additional basis for judging relatedness. Classical taxonomy seeks both to group species that share similar features and to describe evolutionary relationships.
- Cladistics uses characteristics present in a common ancestor and derived characteristics unique to a given taxon to determine the relatedness. Closely related species are assumed to have more derived characteristics in common. Cladistics, then, is concerned with establishing a line of descent.

THINK ABOUT IT

1. What was Linnaeus's contribution to biology?

2. On official documents, you are asked to give your first and last names. Which name is analogous to a genus, and which to a species name?

3. Organize the following categories from most inclusive to least related: genus, domain, order, phylum, species, kingdom, class.

4. List three types of characters used to group organisms in classical taxonomy.

5. What is an analogy? Is an analogy more likely to be found in closely or distantly related species?

6. What does cladistics concern itself with?

7. Imagine that the following are all species of small mammals recently discovered in the Amazonian forests. They all bear an overall similarity in body shape. Your job is to build a phylogenetic tree that best represents their evolutionary history.

    a. Which characters are likely to be ancestral?

    b. Which are derived characters?

    c. Which two species branched most recently (i.e., share the greatest number of characteristics)?

Character/Species	Georgia	Lark	Quinn	Mimi	Borg
Six toes	X	X		X	X
Blue eyes	X	X	X	X	X
Whiskers	X			X	
Meat-eating	X	X	X	X	X

Cool Fact: When One Species Is Really Two

In Africa, elephants live on both the savanna and in the forest. In Asia, elephants live in the forest. Now Asian elephants are distinctly different from African elephants; they look different, they behave differently, they use a different habitat (mostly), and they are genetically distinct. But the African elephants on the savanna also look pretty different and behave differently than their forest cousins, so are they really one species that exploits multiple habitats?

Not so, says Alfred Roca and colleagues. They recently surveyed the sequence of four genes in 195 African elephants and 7 Asian elephants and found that, genetically, the African forest elephants are as almost as different from the savanna elephants as African elephants are from Asian elephants. Furthermore, it didn't matter where in Africa the elephant population was located; savanna elephants from widely separated geographic regions were more similar to each other, genetically, than to the forest elephants within the same location. This genetic evidence is supported by results from classical, morphological studies that have also suggested that the African elephant population is really two species—*Loxodonta africana*, the savanna elephant, and *Loxodonta cyclotis,* the forest elephant.

Does it make a difference to have two species instead of one? Yes, especially if you are the wildlife manager responsible for developing conservation strategies. Acknowledging that not all African elephants are the same has a big impact upon how you plan to set up reserves, breed elephants in captivity, and make judgments about how this endangered species is "rebounding."

Chapter 18

TESTING, TESTING

Make a web: speciation, allopatric, sympatric, hybrid inviability, temporal isolating mechanism, geographic separation

Make a flowchart: Demonstrate cladogenesis by punctuated equilibrium.

Vocabulary Review

Place the following terms in the appropriate columns of the table. Some terms may be used in more than one column.

adaptive radiation
allopatric
cladistics
Linnaeus
population
phylum
species
sympatric
systematics
taxonomic system

Speciation	Categorization

PRACTICE TEST

After you have finished studying this chapter, close your books, grab a pencil, and spend the next 15 to 20 minutes completing this practice test.

Compare and Contrast

For each of the following pairs of terms, write one sentence describing their similarities and one sentence explaining their major differences. In writing about similarities, start the sentence with "Both . . . "; for differences, start with "However . . . "

cladogenesis/anagenesis

extrinsic isolating mechanism/intrinsic isolating mechanism

phylum/kingdom

analogous/homologous

Matching

Match the following terms with their description. Each choice may be used once, more than once, or not at all.

a. systematics
b. punctuated equilibrium
c. sympatric speciation
d. allopatric speciation
e. adaptive radiation

____Speciation takes place in brief bursts, separated by period of stasis
____Occurs due to geographic separation
____Occurs in the absence of geographic separation
____Studying evolutionary history of groups of organisms
____Several new species descended from one species and becoming more specialized to specific habitats

Short Answer

1. What is the difference between sympatric and allopatric speciation?

2. Describe the process of adaptive radiation.

3. What is the major difference between the gradualism and punctuated equilibrium models?

4. In cladistic analysis, what determines the degree of relatedness among species?

Multiple Choice

Circle the letter that best answers the question.

1. The biological species concept fails to recognize species of

   a. primates.
   b. bacteria.
   c. vertebrates.
   d. trees.
   e. fish.

2. Two organisms of different species mate and produce and offspring. This offspring cannot mate with its siblings, nor with members of either parental species. This is an example of

   a. hybrid infertility.
   b. gametic isolation.
   c. behavioral isolation.
   d. ecological isolation.
   e. temporal isolation.

3. In the Galapagos finches, adaptive radiation occurred_________ islands whereas allopatric speciation occurred__________ islands.

   a. within; on large
   b. between; within
   c. on large; on small
   d. within; between
   e. on small; on large

4. The most inclusive category is

   a. kingdom.
   b. genus.
   c. phylum.
   d. domain.
   e. order.

5. Convergent evolution can result in species with

    a. homologous structures.
    b. systematic structures.
    c. analogous structures.
    d. derived characters.
    e. None of the above.

6. Which of the following is used to determine whether two populations belong to the same species?

    a. ability to interbreed
    b. color
    c. size range
    d. habitat choice
    e. name

7. Two individuals from different populations attempt to mate but are unable to successfully coordinate the mating dance. This is an example of a(n)________ isolating mechanism.

    a. ecological
    b. temporal
    c. behavioral
    d. mechanical
    e. gametic

8. Speciation as a small series of changes that accumulate over time is known as

    a. punctuated equilibrium
    b. cladistics
    c. gradualism
    d. stasis
    e. compounded interest

9. Classification of species based on their relatedness is known as

    a. eurythmics.
    b. systematics.
    c. asthmatics.
    d. mathematics.
    e. aeronautics.

10. You are doing a cladistic analysis on a group of five species. How many would you expect to share the most ancient derived trait?

    a. 1
    b. 2
    c. 3
    d. 4
    e. 5

11. You now begin the task of grouping your many Neko fossils. Can you determine which examples are from the same species by using the biological species concept?

    a. No, because the specimens are sympatric.
    b. No, because the specimens are allopatric.
    c. No, because you cannot determine the potential for interbreeding between samples.
    d. Yes, because of the similarities of the skeleton to other modern species.
    e. Yes, because as long as the samples are biological, they will satisfy the concept definition.

12. Plants are much more likely than animals to form hybrids. Most of these hybrids are sterile, but occasionally, these individuals can form the basis for a new species. This is because plants, unlike animals, are capable of surviving

    a. polyploidy.
    b. aploidy.
    c. punctuated equilibrium.
    d. convergent evolution.
    e. harsh winters.

BUT, WHAT'S IT ALL ABOUT?

Here's a question to help you pull together what you've learned so far using this text. If you don't remember how to attack this type of question, check out the example in Chapter 1.

If we look at the phylogenetic trees within this chapter in the textbook, we see a trend from comparatively simple organisms on the leftmost branches to more complex organisms on the right branches. Why does the process of evolution lead to more and more complexity?

What do I do now?
Remember the drill—decide what the question is asking you to do, collect your evidence from this chapter (and the others you've studied) and write!

19

CHAPTER 19

A Slow Unfolding: The History of Life on Earth

Basic Concepts

- The geologic time scale begins 4.6 billion years ago with Earth's formation and is divided into time periods determined by the type of fossils that have accumulated.
- Life, defined as self-replicating molecules, appeared on Earth from 3.5 to 4 billion years ago. All life can be traced to a universal ancestor.
- The Cambrian Explosion marks a remarkable increase in the diversity of animal forms and an increase in the rate of evolution of new animal forms.
- The human family tree begins as chimpanzees and hominids diverge from a common primate ancestor. The most commonly accepted hypothesis, the "out-of-Africa" hypothesis, proposes that modern humans evolved in Africa and then migrated over the globe.

Tracing the History of Life on Earth

- The geologic time scale is divided into eras, which are subdivided into periods, which are divided into epochs. The boundaries between the time periods are marked by major changes in the fossil record.
- "Life" probably began around deep-sea vents or hot-spring pools as organic molecules reacted with each other to form more complex molecules.
- Many scientists believe that RNA molecules were the first self-replicating molecules, because RNA can serve as a template for the transfer of information as well as a catalyst for chemical reactions.

THINK ABOUT IT

1. Approximately how old is the Earth?

2. Rank these time measurements from smallest to largest: millennium, period, era, epoch, decade.

3. What is thought to be responsible for the Cretaceous Extinction?

4. Why do extinctions seem to be followed by the appearance of new species?

5. Assign the letter A (Archaea), B (Bacteria), or E (Eukarya) to each of the following:

_____ plants
_____ E. coli
_____ single-celled in sulfur hot springs
_____ fungi
_____ roaches

6. What critical type of molecule must have been formed in order for life to evolve?

7. Why is RNA an attractive candidate for the first genetic material?

8. List three possible original sources for the raw materials of life.

9. Which of these sources seems most likely to you? Why?

The Tree of Life

- All living things on Earth descended from a "universal ancestor." Three branches sprout from this ancestor—the domains Bacteria, Archaea, and Eukarya.
- The Precambrian period, beginning with Earth's formation, is defined by the appearance of life in the form of bacteria. These bacteria are responsible for producing oxygen through the reactions of photosynthesis.
- Animals first appeared in the Precambrian era, about 600 Mya; but the mass extinctions that accompanied the "oxygen holocaust" opened up niches that made possible the diversification of life forms, in turn making the Cambrian Explosion possible.

THINK ABOUT IT

1. Of the three domains of life, which are most closely related?

2. Use the following table to list the three domains of life and the types of cells in each.

Domain	Cell Type

3. Which kingdom contains species that gave rise to the other kingdoms?

4. What were the first organisms on Earth? When did they first appear?

5. How was the first oxygen produced on Earth? What are some of the benefits of oxygen production?

The Cambrian Explosion

- Plants were the first organisms to make the transition from the seas to the land, about 460 Mya. These plants adapted to the drier land environment by developing a waxy cuticle to minimize water loss and a supportive vascular system to counter the effect of gravity. The major adaptation, however, was shifting the maturation of the embryos to a site within the parent instead of outside in the water.
- Animals followed plants onto land about 400 Mya; the first were wingless insects, followed by the lobe-finned fishes, which are the ancestors of all tetrapod vertebrates. Another critical adaptation to live on land was the appearance of the amniotic egg—a structure that surrounds the developing embryo with padding, food, and a hard, protective covering.
- Mammals appeared about 220 Mya, during the time of the dinosaurs. Mammalian evolution did not take off until after the sudden extinction of the dinosaurs opened niches, about 100 Mya.

THINK ABOUT IT

1. When was the Cambrian Explosion? What was its significance?

2. What is the connection between increased amounts of oxygen in the atmosphere and the appearance of animals on Earth?

3. What were the first multicellular terrestrial organisms?

4. What three problems had to be overcome by the first land plants? What were the solutions?

Problem	Solution

5. What reproductive characteristics separates all plants from their algal relatives?

6. Which would you expect to grow taller, a moss (bryophyte) or a fern (seedless vascular)?

7. Rank the following adaptations from oldest to most recent: vascular system, waxy cuticle, flowers, seeds, photosynthesis.

8. What was the purpose of the exoskeleton of the first land animals?

9. Place the following life forms in the correct order of their appearance on land: insects, reptiles, plants, mammals, amphibians.

10. What is an amniotic egg, and what was its evolutionary impact?

11. How did mammals move away from the amniotic egg?

Human Evolution

- Primates, the ancestors of the modern chimpanzees and humans, appeared 60 Mya. Tracing the evolution of modern humans depends upon interpreting structural and DNA clues present in fossil skeletons.
- Because evidence (both fossil and DNA) of human evolution is incomplete, several possible family trees have been proposed. Most scientists agree that humans evolved from African hominids, abandoning a tree-dwelling lifestyle and adopting an upright stance and bipedalism.

THINK ABOUT IT

1. Where have the oldest hominid fossils been found?

2. Suppose you and a Neanderthal relative sat down to make a family tree. Are Neanderthals a different genus, or species, than humans?

Cool Fact: More than Just Time

Most people understand the premise behind radiocarbon dating; because radioactive elements "decay" at a predictable rate, we can use the ratio of radioactive to nonradioactive elements to estimate the age of ancient things—mummies, fossils, the Earth. But the Earth keeps within its crust more than a calendar. The presence of other elements and molecules can provide clues to ancient climate as well.

In this age of "global warming," one question that has been asked repeatedly is whether the warming trend currently observed is truly a problem or just the reflection of a natural warming and cooling cycle. One way to answer this question would be to determine if the amount of "greenhouse gas," specifically CO_2, in the atmosphere has dramatically increased over levels present in the geological past. CO_2 is marginally soluble in water, and it can be trapped in ice. By boring into the polar ice and measuring the amount of CO_2 trapped within the ice at increasing depths, researchers are able to estimate the amount of CO_2 that must have existed in the ancient atmosphere.

Unfortunately, CO_2 is not trapped uniformly, so different samples within a region can lead to different predictions of ancient CO_2 levels. But over the last 5 to 10 years, as a result of many independent studies, scientists have accumulated a great deal of evidence supporting.

The data suggest that the current warming trend does not reflect just the Earth's normal warming and cooling cycle. Instead, the warming trend is mostly due to human activities, such as deforestation and the burning of fossil fuels. So the Earth not only keeps a record of where life has been but also makes it possible for us to predict its future course.

TESTING, TESTING

Make a web: *Homo*, Ardepithecus, Australopithecus, primate, *Homo sapiens*, *Homo antecessor*, hominid, *Homo neanderthalensis*

Make a flowchart: Cambrian Explosion, Cretaceous Extinction, Permian Extinction, dinosaurs, mammals

Vocabulary Review

Place the following terms in the appropriate columns of the table. Some terms may be used in more than one column.

amniotic egg   Cambrian   epoch   Permian   Protista

bryophyte   continental drift   era   Primates   seedless vascular plant

Geological time	Family Trees

PRACTICE TEST

After you have finished studying this chapter, close your books, grab a pencil, and spend the next 15 to 20 minutes completing this practice test.

Compare and Contrast

For each of the following pairs of terms, write one sentence describing their similarities and one sentence explaining their major differences. In writing about similarities, start the sentence with "Both . . ."; for differences, start with "However . . ."

angiosperm/gymnosperm

Archaea/Bacteria

epoch/era

domain/phylum

Matching

Match the following terms with their description. Each choice may be used once, more than once, or not at all.

a. lemur
b. accretion
c. Archaea
d. bryophyte
e. Bacteria

____ Modern-day descendant of early primates
____ Mosses (seedless, nonvascular plants)
____ Earliest organisms
____ Clumping of cosmic particles into a larger object
____ A prokaryotic domain of life

Short Answer

1. Where did early human evolution take place?

2. What two adaptations allowed plants to colonize land?

3. What is the relationship between the terms "continental drift" and "adaptive radiation"?

4. What is significant about the Permian Extinction?

Multiple Choice

Circle the letter that best answers the question.

1. Based on structural similarities, you conclude that your fossil discovery, *Neko borgus*, was a type of early mammal. *Neko borgus* would therefore have been preceded onto land by all of the following organisms except:

   a. reptiles.
   b. fish.
   c. plants.
   d. humans.
   e. All of these organisms would have preceded *N. borgus* onto land.

2. Amphibians

   a. were the first animals to emerge on land.
   b. lay amniotic eggs.
   c. evolved before reptiles.
   d. live on both land and water.
   e. c and d

3. Which of the following is not a special adaptation of plants to land?

   a. waxy cuticle
   b. vascular system
   c. photosynthesis
   d. seeds
   e. flowers

4. Human beings

   a. are mammals.
   b. evolved from the primate line.
   c. evolved in Africa.
   d. descended from tree-dwelling ancestors.
   e. All of the above.

5. What advantage did the hard shells of arthropods provide during the move onto land?

   a. swimming ability
   b. temperature regulation
   c. desiccation (drying) prevention
   d. increased enzymatic rate
   e. no need to exercise to maintain a "hard body"

6. The domains of life are

   a. Prokarya, Eukarya
   b. Bacteria, Eukarya, Archaea
   c. Plantae, Animalia, Bacteria
   d. Bacteria, Eukarya
   e. None of the above.

7. Lucy *(Australopithecus afarensis)* had a cranial capacity most similar to that of a

   a. gibbon
   b. orangutan
   c. siamang
   d. gorilla
   e. chimpanzee

8. Which of the following is the most similar to humans in terms of adaptations?

   a. gibbon
   b. gorilla
   c. chimp
   d. orangutan
   e. siamang

9. Photosynthetic organisms produce ______, which protects terrestrial organisms from ________.

   a. oxygen; UV irradiation
   b. oxygen; toxins
   c. CO_2; UV irradiation
   d. mitochondria; toxins
   e. CO_2; water loss

10. Which is the most successful plant group (in number of species) that exists today?

    a. bryophytes
    b. ferns
    c. gymnosperms
    d. angiosperms
    e. b and d

11. Based on structural similarities, you conclude that your fossil discovery, *Neko borgus*, was a type of early mammal. What domain would you place this species in?

    a. Archaea
    b. Bacteria
    c. Eukarya
    d. Plantae
    e. Protista

12. The "missing link" in the understanding of molecular evolution on early Earth was the lack of an assembly mechanism for the information storage molecules, DNA and RNA. This link was found with the discovery of the enzyme ribozyme, which is actually a type of

    a. protein.
    b. carbohydrate.
    c. DNA.
    d. RNA.
    e. ATP.

BUT, WHAT'S IT ALL ABOUT?

Here's a question to help you pull together what you've learned so far using this text. If you don't remember how to attack this type of question, check out the example in Chapter 1.

The time scale for life on Earth (Figure 19.3 in the text) shows that life on Earth was rather slow to get started but rapidly diversified beginning 544 Mya during the Cambrian explosion. One can argue that evolution can be easily compared to forming a self-replicating living organism. What challenges must be conquered to create that first living organism?

What do I do now?

Remember the drill—decide what the question is asking you to do, collect your evidence from this chapter (and the others you've studied) and write!

20

CHAPTER 20

Viruses, Bacteria, Archaea, and Protists: The Diversity of Life 1

Basic Concepts

- Although living organisms show extreme diversity of form and function, scientists attempt to group similar organisms into one of three domains: Bacteria, Archaea, and Eukarya.
- Because viruses can replicate only when they take over the replication machinery of another cell, some scientists do not consider them "living" organisms. Thus they define their own category—not "living," not abiotic, and generally bothersome to other organisms.
- Domain Bacteria consist of single-cell, independent-living, haploid organisms that lack all intracellular organelles including a nucleus. Bacteria, however, exist in every environment and are master decomposer organisms.
- Similar to Bacteria but genetically distinct, Archaea are single-cell organisms that inhabit the most extreme environments on Earth. Some scientists argue that Archaea may also be the oldest organisms on the evolutionary tree.
- The kingdom Protista is comprised of organisms that, while eukaryotic, do not have all of the distinguishing characteristics of plants, animals, or fungi.

Viruses

- Viruses are little more than nucleic acids, DNA or RNA, encased in a protein coat.
- Viruses inject their nucleic acids into other cells and commandeer the host cell's replication machinery to make more viruses.
- Viruses infect virtually all other living things. Humans respond to viral infection through an immune system that functions to eliminate the viral invaders.

THINK ABOUT IT

1. Is a virus a cell?

2. How does a virus reproduce?

3. Describe the principle behind vaccination. What vaccinations have you had?

Domain Bacteria

- Bacteria are single-celled, asexually reproducing, free-living organisms possessing a single chromosome. Bacteria lack intracellular organelles and a cytoskeleton.
- Bacteria are the primary decomposers in the biosphere. They exist everywhere, including within other organisms. As residents of other organisms, Bacteria may also fix nitrogen or produce useful sugars and vitamins.
- Bacteria may also cause disease by producing toxins, or fight disease by producing toxins against other microorganisms.

THINK ABOUT IT

1. Are all Bacteria harmful to humans?

2. What beneficial roles do Bacteria play in the environment?

3. Bacteria are quite diverse. What five features do they all share?

   a.

   b.

   c.

   d.

   e.

4. Why are antibiotics more effective against Bacteria than viruses?

Domain Archaea

- Archaea may represent the first type of organism to have evolved on Earth.
- Archaea are similar to Bacteria in that they are free-living, unicellular organisms lacking intracellular organelles. However, Archaea are genetically distinct from both Bacteria and eukaryotic organisms.
- Archaea are frequently found in environments characterized by extremes of temperature, pH, pressure, and salinity.

THINK ABOUT IT

1. What kind of environment does an extremophile live in? If there was life on Venus, what do you think it would look like? (Keep in mind that Venus is a lot closer to the Sun than to Earth.)

2. Evolutionarily, where do Archaea seem to fit in to the scheme of things?

3. Why has it taken so long to recognize the Archaea as a separate domain?

4. Aside from laundry detergent, what other uses can you imagine for Archaea?

Kingdom Protista

- The kingdom Protista is defined by what its members are not—mostly single-celled organisms that are not plants, not fungi, and not animals.
- Protista were the first to develop sexual reproduction.
- Protista acquire energy either through photosynthesis, consumption of other organisms, or decomposition.

1. What do all protists have in common? List 5 common attributes.

   a.

   b.

   c.

   d.

   e.

2. When did protists evolve?

3. What kind of protist

   a. extends filaments into the food source and starts breaking down or decomposing it from within?

   b. demonstrates colonial multicellularity, growing to a very large size as it floats in the oceans harvesting sunlight for energy?

4. Use the chart below to describe the life cycle of the cellular slime mold, *Dictyostelium discoideum*.

Life Stage:	Activities:	Cell Types:
Feeding		
Traveling		
Reproduction		

Cool Fact: The Mystery of the Archaea

Just where in the family tree do the Archaea belong? Some of their genes resemble those of Bacteria, some of eukaryotic cells, but most are completely unique. What is the evidence that places Archaea in its own domain?

Work by Carl Woose and colleagues using the sequences of rRNA genes suggested that Bacteria and Archaea arose from a common ancestor, and that the domain Eukarya evolved from Archaea. Woose chose to study rRNA genes because ribosomes are absolutely essential for life and would be expected to accumulate mutations very slowly. However, as research studies explored sequences of other genes, the order of domains at the base of the family tree changed depending on the gene sequence used. Some analysis showed more genetic similarities between eukaryotes and Bacteria than between eukaryotes and their presumed ancestors, Archaea.

Why such diversity? Possibly because microbes are notorious gene swappers, collecting genes from their food or by lateral transfer from neighboring organisms. All of this sharing, stealing, and swapping may be partially responsible for the universality of the DNA genetic code, because non-DNA genes would be rapidly lost. While it's a boon for evolution, freely "moving" DNA makes it difficult to pin down the order of evolution at the base of the family tree. Whether Archaea merit a separate domain—or represent a branch of the Bacteria domain—remains an open question in many scientists' minds.

Chapter 20

TESTING, TESTING

Create a web: The domains Archaea, Eukarya, and Bacteria; single-celled; sexual reproduction; nucleus; asexual reproduction; cells

Vocabulary Review

Place the following terms in the appropriate columns of the table. Some terms may be used in more than one column.

algae	Eukaryotic	Giardia	photoautotrophy
antibiotic	extremophile	haploid	phytoplankton
asexual reproduction	fungi-like	pathogenic	prokaryotic

Archaea	Bacteria	Protista

PRACTICE TEST

After you have finished studying this chapter, close your books, grab a pencil, and spend the next 15 to 20 minutes completing this practice test.

Compare and Contrast

Archaea/Bacteria Both are single-celled prokaryotes. However Archaea are more likely to be found in extreme environments, high temperatures or high salt content, for example, while Bacteria are not.

autotrophs/heterotrophs

phytoplankton/zooplankton

viruses/protists

Matching

Match the following terms with their description. Each choice may be used once, more than once, or not at all.

a. algae
b. virus
c. protozoan
d. extremophiles
e. Bacteria

____ Non-cellular infectious agent
____ Animal-like protist

____ Aquatic protists
____ Thrives in high heat environment
____Prokaryotic single-celled organisms

Short Answer

1. Which domain is most similar to the Archaea, and why?

2. What are phytoplankton?

3. What is penicillin?

4. What is the function of pseudopodia?

Multiple Choice

Circle the letter that best answers the question.

1. Which of the following are the domains of life?
   a. Eukarya, Fungi, Algae
   b. Eukarya, Viruses, Prokaryotes
   c. Eukaryotes, Prokaryotes, Plants
   d. Eukarya, Archaea, Bacteria
   e. Plants, Animals, Fungi

2. Photosynthetic organisms include
   a. protists.
   b. fungi.
   c. Bacteria.
   d. yeasts.
   e. a and c

3. Which of the following is a protist?
   a. mushroom
   b. whale
   c. worm
   d. algae
   e. fish

4. Sewage treatment plants use

    a. Bacteria.
    b. Archaea.
    c. yeasts.
    d. a and b
    e. All of the above.

5. Bacteria are

    a. always harmful to humans.
    b. always beneficial to humans.
    c. very large viruses.
    d. generally haploid, unicellular organisms.
    e. never found in nature.

6. Viruses are part of the

    a. domain Bacteria.
    b. domain Archaea.
    c. kingdom Protista.
    d. kingdom Viridia.
    e. None of the above.

7. Are viruses alive?

    a. No, their lack of replicative ability prevents classification as alive.
    b. No, they don't contain genetic material.
    c. Yes, all pathogens are living organisms.
    d. Yes, they use a protein coat instead of a cell plasma membrane.
    e. Yes, they are capable of very sophisticated behavior and therefore, must be alive.

8. Bacteria within the human body

    a. don't exist.
    b. may cause disease.
    c. may help in digestion of food.
    d. both A and B
    e. both B and C

9. Microscopic organisms (microbes)

    a. produce more than half of the world's oxygen
    b. provide sources of nitrogen for plants to use
    c. act as decomposers of organic material
    d. All of these.

10. Instead of sterilizing your water bottles after your weekend camping trip as you usually do, you decide to put the few remaining drops of water on a microscope slide. You observe a diverse group of single-celled creatures. After consulting a few books, you are pretty certain the organisms are NOT Archaea. Why not?

    a. Archaea cannot exist in an oxygen atmosphere.
    b. ARCHAEA are unlikely to be found in areas where you would go for a weekend camping trip.
    c. Archaea are too large to fit on a microscope slide.
    d. Archaea are never found in water.

11. You try to determine more about your mysterious creatures. Which of the following attributes will tell you definitively whether they are prokarya or Eukarya?

    a. photosynthesis
    b. single-celled
    c. sexual reproduction
    d. presence of DNA

12. Suppose you determine that your creatures are eukaryotic and acquire energy by decomposing organic matter. What type of protists would they be?

    a. fungi-like
    b. animal-like
    c. plant-like
    d. Bacteria-like

BUT, WHAT'S IT ALL ABOUT?

Here's a question to help you pull together what you've learned so far using this text. If you don't remember how to attack this type of question, check out the example in Chapter 1.

Members of the Kingdom Protista seem to defy the orderly patterns of evolution seen in Chapter 19 where a simple ancestral organism evolves into organisms of increasing complexity by adding more features. The Protists, however, seem to be a collection of organisms of diverse size, shape, and lifestyle, some of which are related to the ancestors of modern plants, others related to the ancestors of modern animals, and still others related to the ancestors of modern Protists. Why do the organisms in this Kingdom persist?

What do I do now?

Remember the drill—decide what the question is asking you to do, collect your evidence from this chapter (and the others you've studied) and write!

21

CHAPTER 21

Fungi and Plants: The Diversity of Life 2

Basic Concepts

- Despite the great diversity in size, shape, and function, all members of Eukarya possess nucleated, diploid cells that also contain other organelles and structural proteins. Members of the domain Eukarya are the only organisms that show true multicellularity. Eukarya is divided into four kingdoms: Protista, Fungi, Plantae, and Animalia.

Kingdom Fungi and Protista

- Members of the kingdom Protista are defined by what they are not—mostly single-celled organisms that are not plants, not fungi, and not animals.
- The fungi are not plants because they are heterotrophic. Fungi are important to the biosphere as decomposers and mineral and water scavengers for other organisms. Some fungi are pathogenic.

THINK ABOUT IT

1. What are the four kingdoms in the domain Eukarya?

2. What is the major difference between fertilization in humans and fusion in fungi?

3. Would you use an antibiotic like penicillin to treat athlete's foot? Why or why not?

4. Label the figure below with the following terms: dikaryotic, diploid, haploid, meiosis.

LIFE CYCLE OF A FUNGUS

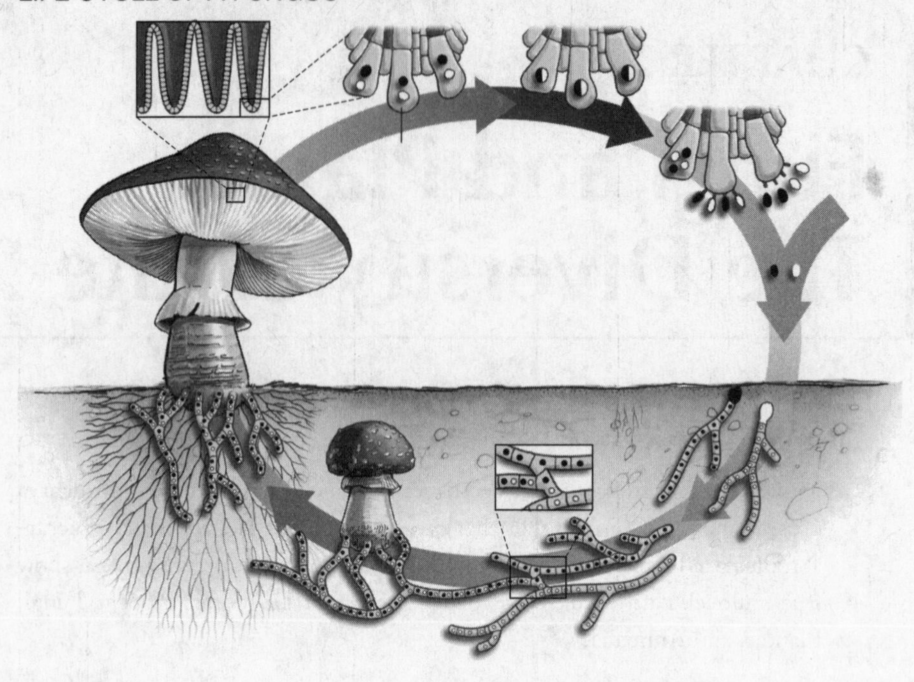

5. What are the two components of lichen, and what role does each play?

Kingdom Plantae

- Members of the kingdom Plantae are mostly sessile, multicellular organisms that can make their own food through the process of photosynthesis. Plants are also characterized by the presence of rigid cell walls, which regulate water uptake, and by a life cycle featuring alternating haploid and diploid generations.
- There are four main categories of plants: bryophytes, seedless vascular plants, gymnosperms, and angiosperms. These categories are defined by the presence or absence of a vascular system and by method of reproduction, either spores or seeds.

THINK ABOUT IT

1. For each of the following plant groups, give a common example and where you have seen it. (Example: angiosperm—an orchid at your neighbor's house)
   bryophyte, seedless vascular plant, gymnosperm, angiosperm

2. What is a seed? Why is seed production found among the most successful plant groups today?

3. Why do plants have flowers?

4. What is the function of endosperm? List the last two endosperm sources that you consumed.

5. Compare and contrast cherries and cucumbers in terms of function.

6. Match the major adaptations of plants to land (multicellularity, flowers, seeds, vascular system) to the labels on the figure below.

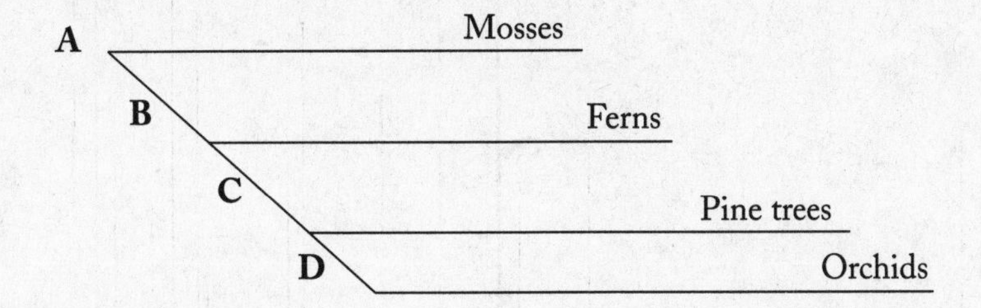

Cool Fact: Fun Fungi?

Is there anything more unassuming than a mushroom? Buried in leaves in the dim light of the woods, squat and brown (mostly), how did anyone decide that these plant-like structures were edible? Most mushrooms, if clean of insect infestation, are safe to eat – whether they are worth eating is a matter of personal taste. A rare few, however, provide an unexpected dividend – visions, perhaps, or even death. Fungi of the genera *Psilocybe* make hallucinogenic alkaloids called psilocybins. These species constitute the "magic mushrooms" that are said to produce "visions of the infinite" by those who consume or smoke the dried products. A vision of infinite darkness might overtake those consuming *Amanita phalloides*, the "angel of death." A rare and beautiful fungi, completely white with a gracefully sloping cap over a delicate "skirt" of membrane, these fungi make phallotoxin and amanitins. Phallotoxin produces gastrointestinal distress, but the amanitins are more toxic, producing renal and kidney damage serious enough to cause death if consumed in high enough quantities. Fortunately, fatal poisonings seem to be very rare because the mushrooms themselves are small, rare, and fragile so it would be difficult for someone to collect enough to produce a deadly delicacy.

Chapter 21

TESTING, TESTING

Create a web: The Domain Eukarya kingdoms Fungi and Plantae; photosynthesis; algae; lichens; gymnosperms; organelles, mycorrhizae, decomposers

Vocabulary Review

Place the following terms in the appropriate columns of the table. Some terms may be used in more than one column.

chloroplast	Eukaryotic	molds	spores
diploid	haploid	rhizoids	vascular system
endosperm	hypha	seeds	yeast

Plants	Fungi

PRACTICE TEST

After you have finished studying this chapter, close your books, grab a pencil, and spend the next 15 to 20 minutes completing this practice test.

Compare and Contrast

For each of the following paired terms, write a sentence of comparison ("Both . . . ") and a sentence of contrast ("However, . . . ").

autotroph/heterotroph

cup fungi/imperfect fungi

sporophyte/gametophyte

angiosperm/gymnosperm

Matching

Match the following terms with their description. Each choice may be used once, more than once, or not at all.

a. gymnosperms
b. mycorrhizae
c. dikaryotic
d. ferns
e. chloroplasts

___ Containing two haploid nuclei
___ Form associations with plant roots
___ Site of photosynthesis
___ Multicellular, vascular system
___ Multicellular, vascular system, seeds

Short Answer

1. What is unusual about the sperm of bryophytes?

2. What is a lichen?

3. Describe two adaptations that gymnosperms have to living on land.

4. What is the function of nectar?

Multiple Choice

Circle the letter that best answers the question.

1. Fungi have formed mutualistic association with all of the following EXCEPT:

   a. protista.
   b. bacteria.
   c. plants.
   d. animals.

2. Which of the following is NOT a classification of fungi?

   a. cup fungi
   b. club fungi
   c. bread molds
   d. perfect fungi

3. Which of the following is NOT shared by all fungi?

   a. hyphae
   b. heterotrophic
   c. mycelium
   d. eukaryotic
   e. All of these are common characteristics of fungi.

4. The dinner conversation with your new sweetheart's family is not going well. To move the discussion along, you start describing the various classifications of fungi.

Your sweetie's obnoxious younger sister challenges you to place the mushrooms in the evening's entrée in the correct category. Without hesitation you (correctly) say,

a. "Imperfect fungi!"
b. "Cup fungi!"
c. "Club fungi!"
d. "Bread mold!"

5. Plants use several methods to attract animals to their reproductive structures. Why do they want these pollinators to carry away pollen after their visit?

a. pollen contains the plant sperm for reproduction.
b. pollen contains the toxic by-products of photosynthesis.
c. fungal plant invaders force the plant to produce pollen to distribute their spores.
d. bacteria tend to infest the plant parts known as pollen and the plant is trying to remove them.

6. All of these are shared common features of plants EXCEPT:

a. photoautotrophic.
b. multicellular.
c. mobile.
d. eukaryotic.

7. A common form of seasonal allergy is the result of the release of pollen by plants. Which of the following would NOT be responsible for your itchy, watery eyes?

a. bryophytes
b. vascular seedless
c. gymnosperms
d. angiosperms
e. both A and B

8. Which of the following correctly matches the classification and the example?

a. bryophyte—mosses
b. seedless vascular—cactus
c. gymnosperm—fern
d. angiosperm—pine tree

9. Which of the following most closely resembles the earliest plants?

a. bryophytes
b. seedless vascular

c. gymnosperms
d. angiosperms
e. None of the above.

10. You find a new plant species with vascular bundles, but no flowers. How would you classify it?

a. bryophyte
b. fern
c. gymnosperm
d. angiosperm
e. b or c

11. The plant life cycle is characterized by alternation of generations. The generation that produces spores is known as the _________________ generation, whereas the one that produces the seeds is the _________________ generation.

a. sporophyte; gametophyte
b. haploid; diploid
c. bryophyte; angiosperm
d. neophyte; gametophyte
e. parental; offspring

12. Your ship crash lands on planet X2-Alpha and all of your medical supplies are destroyed. Luckily, X2-Alpha has many of the flora common to Earth. Which of the groups below would be a good likely source for potent antibiotics like penicillin?

a. bread molds
b. club fungi
c. cup fungi
d. animal-like protista
e. archaea

13. Being dikaryotic means

a. cells can be diploid while having haploid nuclei.
b. having both a sporophyte and gametophyte generation.
c. having both flowers and seeds.
d. never having to say you're sorry.

BUT, WHAT'S IT ALL ABOUT?

Here's a question to help you pull together what you've learned so far using this text. If you don't remember how to attack this type of question, check out the example in Chapter 1.

As evolutionary adaptations go, the development of angiosperms seems like a risky proposition. These plants depend upon the existence of some other organism in order to complete their life cycle. Early angiosperms did not "know" that animals existed that would disperse their pollen or seeds, so how could such a dependency develop?

What do I do now?

Remember the drill – decide what the question is asking you to do, collect your evidence from this chapter (and the others you've studied) and write!

22

CHAPTER 22

Animals: The Diversity of Life 3

Basic Concepts

- Animals are a remarkably diverse collection of multicellular, heterotrophic organisms. The single feature uniquely defining all animals is that they pass through a blastula stage during development.
- The most primitive animals lack a central body cavity, a coelom. Sponges (Porifera) and jellyfish (Cnidaria) are examples of noncoelomate animals.
- More complex animals that possess a coelom can be further differentiated by the orientation of their nerve cord. Protostomes have a ventral nerve cord, while deuterostomes have a dorsal nerve cord. Protostomes and deuterostomes define two branches of equally complex animals on the family tree.

Defining an Animal

- Animals are classified according to body plan into phyla (singular, phylum). Body plans become increasingly complex as we move up the family tree because of the addition of body features.
- The most primitive group of animals lacks a body plan—the sponges (Porifera) do not have tissues, symmetry, or a body cavity.
- More complex animal bodies have tissues organized into organs and show some type of body symmetry, either radial or bilateral.
- The most complex animals possess an internal body cavity, a coelom, that provides protection to internal organs and flexibility to the body.
- Coelomate animals are further categorized based on whether the notochord orients to the ventral or dorsal side of the body.

THINK ABOUT IT

1. Why are sponges still classified as animals if they lack a body plan?

2. Are deuterostomes more complex animals than protostomes? Why or why not?

3. For the following diagram, match the phyla to their appropriate boxes: Cnidaria, Porifera, Mollusca, Echinodermata, Annelida, Arthropoda, Chordata, Platyhelminthes, Nematoda.

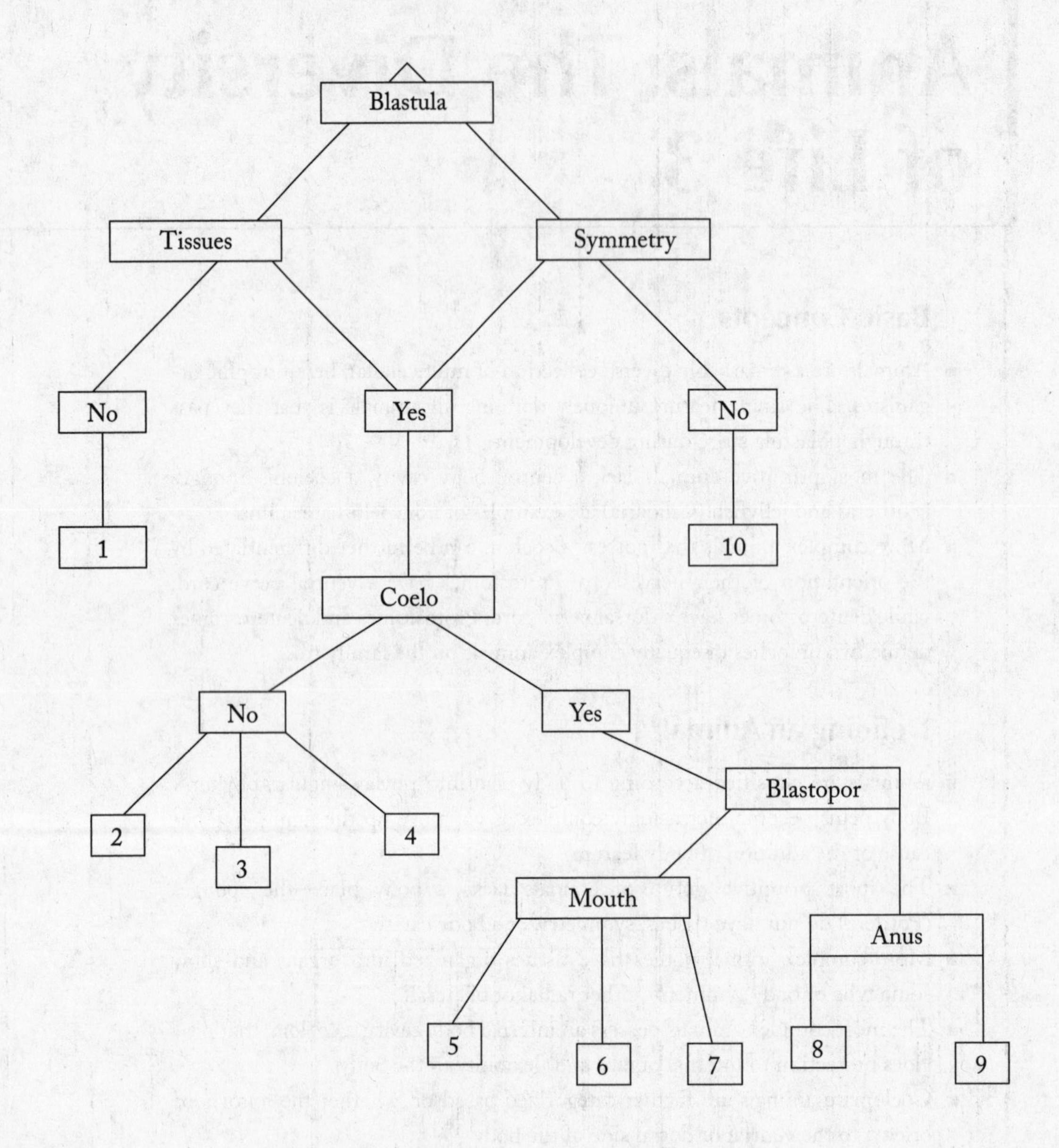

Noncoelomates: Phyla Porifera and Cnidaria

- Sponges are the odd members of the animal kingdom, not even possessing tissues; but a sponge is still a cooperative entity. The independent cells work together to feed by filtering the water that passes through the sponge.
- Cnidarians—jellyfish, corals, anemones, and hydrozoans—possess a sack-like body with a single opening to a gastrovascular cavity that carries out both digestion and nutrient transport.
- Cnidarians have rudimentary nervous and muscle systems that allow the animals to move by expelling a jet of water.

THINK ABOUT IT

1. Lacking any body plan, tissues, organs, or other features of higher organisms, how do the inner cells of a sponge get the food and oxygen they need?

2. Jellyfish are squishy and mobile; corals are hard and sessile. How can these organisms be part of the same phylum (what feature do they share)?

3. What are two major differences between Cnidarians and Porifera? (What is the defining characteristic of Cnidaria?)

Protostomes

- All protostomes possess tissues organized into organs, including nerve cells, bilateral symmetry, and a layer of tissue in the embryo, called the mesoderm, which allows for the development of more complex tissues.
- Flatworms (Platyhelminthes) are the most primitive protostomes, lacking a coelom but possessing a collection of nerve cells and primitive eyes at one end, defining a "head." Most flatworms are parasitic species.
- Phylum Nematoda (Roundworms) are the first protostomes to have a coelom, so their bodies have two openings (mouth and anus). Most roundworms are also parasitic species.
- With the phylum Mollusca we first see a true circulatory system, an open system in which blood is pumped into the tissues and diffuses back to the veins, as well as a stomach and kidney. One group of molluscs (the cephalopods) show signs of intelligence, indications of a developed nervous system. The one feature shared by all molluscs is the mantle, a protective tissue layer on the upper surface of the body.
- Body segmentation appears in the phylum Annelida (segmented worms). Segmentation allows parts of the body to function independently, giving flexibility and strength to the body plan.
- The most numerous protostomes, and most numerous types of animals, are the members of phylum Arthropoda, which includes 1,000,000 species of insects. All arthropods are characterized by an exoskeleton and paired, jointed appendages. In addition to the insects (subphylum Uniramia), other members of this very large phylum include spiders, ticks, and horseshoe crabs (subphylum Chelicerata); and crabs, lobsters, and shrimp (subphylum Crustacea).

THINK ABOUT IT

1. Organization of tissues into organs increases the complexity and functionality of the organism. Why do the protostomes have organs but the Cnidarians and Poriferans do not?

2. Why are flatworms flat?

3. How do arthropods grow when their bodies are covered by an exoskeleton?

4. We first see segmentation as a body feature of annelid worms. Is this a feature shared by more complex protostomes?

5. Survival is closely linked to the ability of an organism to secure enough food. How have insects been able to ensure a food supply?

6. What does it mean to be triploblastic?

7. By now, you may have learned more about worms than you ever cared to know. Demonstrate your expertise by labeling each of the following as characterizing flatworms (F), segmented worms (S), or roundworms (R):

   ___ body segmentation
   ___ hookworms
   ___ hermaphroditic
   ___ pseudocoel
   ___ parasitic tapeworms

8. Place each of the following organisms in the appropriate category: oysters, squid, octopus, clams, mussels, snails, slugs, nautilus.

Cephalopod	Gastropod	Bivalve

Deuterostomes

- The Echinodermata appear to be a "throwback" to the more primitive protostomes; sea stars lack a brain, have primitive eyespots and radial symmetry, and can regenerate from part of the central disc and an arm. But echinoderm embryos have bilateral symmetry, a feature that makes these organisms more developed than Cnidarians.

- Vertebrate animals, including humans, are members of the phylum Chordata. All chordates possess a notochord, a dorsal nerve cord, pharyngeal slits, and a postanal tail, although some members of the phylum, like us, show some of these features only during embryonic development.
- Fifty percent of the chordate species are fish; and among fish, the most common species are ray-finned, bony fishes. However, it was the lobe-finned fishes that served as the ancestor of all land vertebrates.
- Amphibians, such as frogs, are terrestrial vertebrates that must spend some part of their life cycle in an aquatic environment.
- Reptiles and birds, as different as they appear, probably evolved from a common, dinosaur ancestor. These two kinds of animals encase their embryos in an amniotic egg, a structure that protects the developing embryo from drying out.
- Only a small number of chordates are mammals, but these animals (like us) have had a profound effect on the Earth. All mammals have mammary glands, providing milk for the young. They can maintain a constant body temperature, which means they can inhabit any environment, including cold ones.
- Three classes of mammals are distinguished by modes of reproduction. Monotremes are egg-layers. In the marsupials, development of the young animal occurs both inside and outside of the mother's body. Young animals develop only within their mother's bodies among the placental mammals, depending on the mother's blood supply for nourishment.

THINK ABOUT IT

1. What is the difference between a vertebrate and an invertebrate animal?

2. Why would the development of jaws in ancient chordates be a significant advantage to the animal?

3. Why does the amniotic egg represent an evolutionary advantage to the reptiles and birds, compared to the eggs of amphibians?

4. Name the two characteristics shared by all mammals and the two shared by *most* mammals.

All mammals (a)

(b)

Most mammals (c)

(d)

5. Being able to maintain a constant body temperature (endothermy) is a great advantage to mammals, but it comes at a cost. What do mammals need to be endothermic?

6. What separates the Vertebrata from the other two subphyla in Chordata, the Cephalochordata and the Urochordata?

Cool Fact: Vertebrate Evolution

If the "signal shift" in vertebrate evolution was the appearance of jaws, how did this great change happen? Evolutionary biologists usually trace the development of new structures by comparing structures in modern animals to fossil structures. But it is also possible to gain information by comparing the development of modern animals, at least one of which represents an animal that is still evolutionarily "primitive." Such is the case in the study of the evolution of vertebrate jaws.

The lamprey is a modern jawless fish, part of an evolutionarily old class of organism. Scientists have used three lines of evidence to determine whether head and pharyngeal structures of the lamprey gave rise to structures in jawed fishes, like sharks. The first line of evidence compares anatomical structures in the heads of lampreys and sharks and identifies homologous structures. Second, because these are both living organisms, scientists can watch the development of larval lampreys and embryonic sharks and determine which cells develop into which structures. Following nerve development specifically, scientists find that the same cranial nerves in both organisms innervate the structures identified as homologous by anatomical studies. Third, when scientists examine the timing and pattern of gene expression in developing organisms, they find the same genes (the Pax family) are expressed in the presumed homologous structures at the same time during development. The convergence of three independent lines of evidence support the hypothesis that jawed animals evolved from jawless ones.

Chapter 22

TESTING, TESTING

Create a web: Include the vertebrate taxa (amphibians, reptiles, birds, fish, mammals) and their adaptations (hair, postanal tails, notochord, gills, wings, feathers).

Vocabulary

Place the following terms in the appropriate columns of the table. Some terms may be used in more than one column.

Arthropoda	Echinodermata	monotreme
bilateral symmetry	hair	nematodes
blastopore	invertebrate	radial symmetry
coelom	mammary glands	tissues
diploblastic	Mollusca	vertebral column

PRACTICE TEST

After you have finished studying this chapter, close your books, grab a pencil, and spend the next 15 to 20 minutes completing this practice test.

Compare and Contrast

For each of the following paired terms, write a sentence of comparison ("Both . . . ") and a sentence of contrast ("However, . . . ").

protostome/deuterostome

roundworms/flatworms

ectothermy/endothermy

triploblastic/diploblastic

Matching

Match the following terms with their description. Each choice may be used once, more than once, or not at all.

a. reptiles
b. Porifera
c. Mollusca
d. marsupials
e. Platyhelminthes

___ Lacking symmetry and tissues
___ Tapeworms and other flatworms
___ Squid
___ Turtles, snakes, and lizards
___ Kangaroos

Short Answer

1. List the four characteristics common to all animals, and explain which one is found only in animals.

2. Define the term *phylum*.

3. What is the defining feature of the Chordata?

4. Give an example of species found in the subphylum Uniramia (Phylum Arthropoda).

Multiple Choice

1. Your friend Joe spent the summer visiting a number of exotic places and came home with a parasitic infection. In fact, he has been diagnosed as suffering from trichinosis. What phylum is the source of this infection?

   a. Annelida
   b. Playhelminthes
   c. Mollusca
   d. Nematoda
   e. Porifera

2. Which of the following is NOT a type of reptile?

   a. snakes and lizards
   b. turtles
   c. amphibians
   d. crocodiles and alligators
   e. dinosaurs

3. Most animal fossils involve only the hardest structures of organisms. Which of the following phyla is/are therefore most likely to leave behind such evidence?

   a. Chordata
   b. Arthropoda
   c. Nematode
   d. Annelida
   e. a and b

4. You take a job as a research assistant at one of the richest fossil fields ever to be discovered. On your first day, you discover a new species, *Neko borgus*, which bears similarities to the modern domesticated cat. *Neko borgus* belongs to which of the following groups?

   a. Chordata
   b. protostomes
   c. coelomates
   d. Arthropoda
   e. a and c

5. What group comprises more than half of the vertebrate species?

   a. fish
   b. amphibians
   c. mammals
   d. birds
   e. reptiles

6. Which of the following is NOT a defining feature of the Arthropoda?

    a. exoskeleton
    b. jointed appendages
    c. notochord
    d. protostome
    e. coelomate

7. What is the fundamental difference between protostomes and deuterostomes?

    a. bilateral symmetry
    b. the fate of the blastopore
    c. the presence of tissues
    d. the presence of a notochord
    e. the presence of a coelom

8. Which of the following does NOT belong to the phylum Mollusca?

    a. gastropods
    b. nematodes
    c. cephalopods
    d. octopus
    e. bivalves

9. Monotremes are

    a. molluscs.
    b. amphibians.
    c. porifera.
    d. mammals.
    e. annelids.

10. Mammals

    a. are endothermic.
    b. are coelomates.
    c. have mammary glands.
    d. are vertebrates.
    e. are all of the above.

11. Which feature(s) would be shared by both a sea star and a human but NOT a cockroach?

    a. both are protostomes
    b. both are deuterostomes
    c. both have a coelom
    d. both a and c
    e. both b and c

12. A friend brings you a jar of seawater collected on a recent vacation at the beach. Within the jar you find a blob of what appears to be living tissue, which you eventually determine to be a new kind of animal. You are confident in your conclusion because it fits all of the characteristics of an animal. All animals have the following characteristics EXCEPT:

    a. symmetry.
    b. cells without cell walls.
    c. the blastula stage of development.
    d. heterotrophy.
    e. multicellularity.

BUT, WHAT'S IT ALL ABOUT?

Here's a question to help you pull together what you've learned so far using this text. If you don't remember how to attack this type of question, check out the example in Chapter 1.

The family tree of the kingdom Animalia groups animals according to increasing complexity of form. Think back to Chapter 1 (re-read the section on the hierarchy of life to remind yourself of that hierarchy) and the hierarchy of life from molecules to biosphere. How does the hierarchy described by phylogenetics fit into this larger hierarchy of living things?

What do I do now?

Remember the drill – decide what the question is asking you to do, collect your evidence from this chapter (and the others you've studied) and write!

CHAPTER 23

An Introduction to Flowering Plants

Basic Concepts

- Plants are vital to all life on Earth. In addition to providing food for virtually all other life forms on the planet, plants provide oxygen.
- Angiosperms, the flowering plants, account for most of the plant species on Earth. They include important food crops and trees as well as the plants commonly called flowers.

The Structure of Flowering Plants

- Plant structures are either roots or shoots. Roots are responsible for absorbing nutrients and water from the soil. Shoots house the photosynthetic and reproductive centers of the plant.
- There are two types of roots. Taproots consist of a large central root with numerous small side roots. Fibrous roots have many equally sized roots covered with root hairs. All roots grow in search of water, which they need to compensate for water lost from leaves during transpiration.
- Shoots consist of leaves, stems, and flowers. The leaf, which is the site of photosynthesis, has structures to gather CO_2 and sunlight. Stems give structure to the plant and serve as the storage site for food. Flowers are reproductive structures.

THINK ABOUT IT

1. Match the groups to their approximate percentage of plant species in the chart below:

gymnosperms, angiosperms, bryophytes, seedless vascular

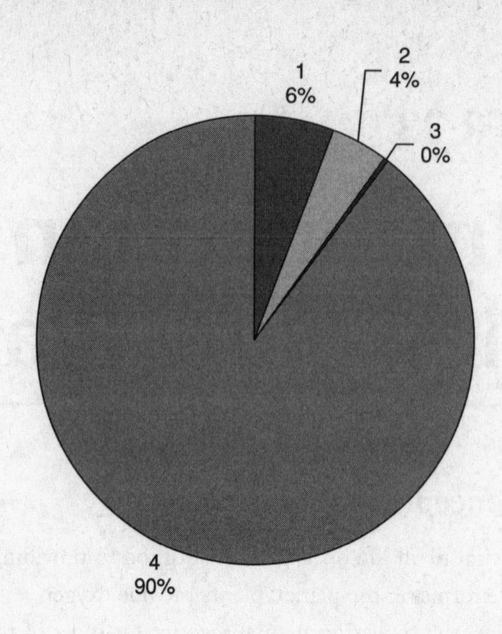

2. Draw a picture of a flowering plant, and label the following structures: roots, root hairs, blade, stem, pedicel, receptacle, sepals, petals, stamens, carpels.

3. Some structures exist within the plant yet are not readily visible. Indicate on your drawing the part of the plant that houses the cuticle, epidermal cells, vascular cells, mesophyll bundles, stomata, outer cortex, pith, filament, anthers, stigma, style, ovary.

4. Mark the following functions with an R or an S, depending on whether it is carried out by the root system or the shoot system.

___ anchoring ___ photosynthesis
___ reproduction ___ water absorption

5. Why do many plants have an extensive root system?

6. Fill in the following table, describing aspects of food and water transport in plants.

	Water Transport	Food Transport
System in Plant		
Direction of Transport		
Cells Involved		

The Function of Flowering Plants

- Flowering plants can reproduce sexually through the production of gametes, sperm, and eggs or asexually through vegetative reproduction (cloning).
  - — Sperm develop within pollen grains; eggs develop within the megaspore in the ovary.
  - — Fertilization results in production of a seed—a structure containing the developing embryo, its food supply, and a protective outer coat.
  - — Reproduction occurs over two generations. The original plant (sporophyte generation) makes diploid spores. These spores (gametophyte generation) make the haploid sperm and eggs.

- Flowering plants transport water from roots to tips through the xylem, a tissue that dies at maturity and supplies both transport and structural functions for the plant.
  - — Nutrients are transported through an adjacent system of cells called the phloem.
  - — Xylem transports water from root to tip, whereas phloem transports food throughout the plant through the action of sieve cells.

- Plants make communication molecules, hormones, and auxins, which direct the plant's growth, defense against predators, and communication with other plants.
  - — Plants grow primarily at their tips (shoots and roots), and growth can continue indefinitely (indeterminate). Some plants can also grow laterally (wood).
  - — Plants protect themselves from predators using physical structures (thorns) as well as chemical defenses.
  - — Auxins mediate the plant's response to its environment, allowing it to grow toward the light or go dormant in the winter.

THINK ABOUT IT

1. Label each of the following terms with an F for female or an M for male.

___ stamen ____ style ____ stigma
___ anther ____ carpel ____ pollen ____ filament

2. Draw an angiosperm tree. Mark a point on it. Now draw the tree after several seasons of growth. Where is the mark?

3. What is phototropism?

4. a. What is a growth response to gravity? How does a plant sense the "direction" of gravity (which way is down)?

   b. What is a growth response to touch?

5. Ragweed normally blooms in the fall, while lettuce blooms in the summer. What is the basis for this difference?

Cool Fact: What Color Is Your Garden?

Flowers delight humans with their wondrous variety of colors and fragrances, but to the plant these sensual delights are all business. Colors primarily attract bird pollinators because birds have a poor sense of taste and smell. That's why bird feeders are brightly colored—those red or yellow housings are signals to birds that food lies within. Flower color can also be a clue to the origin of a plant. No native European plant is pollinated by birds, so there are no red flowers among those plants. Some insects, like bees, respond to color—but not colors visible to our senses. Bees detect color patterns in the UV radiation range, so the yellow and purple of pansies look very different to a bee's eyes.

Fragrance primarily attracts insects, so flowers with insect pollinators may be highly scented but have boring blossoms. Flowers pollinated by moths are generally white and open at night. The carrion flower would hardly be recognized as a flower by humans, and it smells like rotting garbage, but it sends out the sweet smell of dinner for a variety of beetles.

Chapter 22

TESTING, TESTING

Create a web: flowers, seeds, vascular, photosynthesis, fern, moss, pine tree, rose

Make a flowchart: Describe the movement of dissolved minerals through the plant.

Vocabulary Review

Place the following terms in the appropriate columns of the table. Some terms may be used in more than one column.

flowers	phloem	pollen	thigmotropism
hormone	photoperiodism	roots	vascular system
megaspore	phototropism	seeds	xylem

Reproduction	Transport	Signals

PRACTICE TEST

After you have finished studying this chapter, close your books, grab a pencil, and spend the next 15 to 20 minutes completing this practice test.

Compare and Contrast

For each of the following paired terms, write a sentence of comparison ("Both . . . ") and a sentence of contrast ("However, . . . ").

angiosperm/gymnosperm

stamen/carpel

root/stem

xylem/phloem

Matching

Match the following terms with their description. Each choice may be used once, more than once, or not at all.

a. the waxy outer layer of the shoot system
b. plant gas exchange system
c. response to touch
d. movement of water through a plant from roots to leaves
e. sperm must swim in order to fertilize egg

_____ Stomata
_____ Transpiration
_____ Cuticle
_____ Thigmotropism
_____ Bryophyte

Short Answer

1. What is the function of the sieve tube and companion cells?

2. Why are plants important to carnivores like boa constrictors?

3. Describe two ways that plants can defend themselves.

4. What is the meaning of the term *deciduous*?

Multiple Choice

Circle the letter that best answers the question.

1. The earliest plants made the transition from
   a. sexual to asexual reproduction.
   b. water to land.
   c. airborne to earthbound.
   d. dinosaur pollinated to soil pollinated.
   e. photosynthetic to chemosynthetic.

2. All plants have which of the following properties?
   a. flowers
   b. seeds
   c. vascular system
   d. waxy cuticle
   e. All of the above.

3. You find a new plant species with vascular bundles, but no flowers. How would you classify it?
   a. bryophyte
   b. fern
   c. gymnosperm
   d. angiosperm
   e. b or c

4. A plant synthesizes sugar in

   a. the phloem.
   b. the xylem.
   c. the roots.
   d. the leaves.
   e. the gametophyte.

5. A sperm of a plant is found in

   a. the carpel.
   b. the pollen.
   c. the seed.
   d. the fruit.
   e. the sporophyte.

6. What part of the plant does fruit develop from?

   a. the ovary within the flower
   b. the seed
   c. the pollen
   d. the petals of the flower
   e. None of the above.

7. Which of the following most closely resembles the earliest plants?

   a. bryophytes
   b. seedless vascular
   c. gymnosperms
   d. angiosperms
   e. None of the above.

8. Which of the following is the least common group of plants?

   a. bryophytes
   b. seedless vascular
   c. gymnosperms
   d. angiosperms
   e. None of the above.

9. Tracheids and vessel elements, two types of cells used by plants for fluid transport, are part of the

   a. stomata.
   b. chloroplasts.
   c. xylem.
   d. phloem.
   e. endosperm.

10. All plants experience

    a. puberty.
    b. primary growth.
    c. secondary growth.
    d. senescence.
    e. All of the above.

11. Burning incense releases ethylene gas. Suppose that after burning part of a special dinner you were cooking for your sweetheart, you decide to burn incense in the kitchen to get rid of the smell. What would you expect to happen to the apples and bananas in the nearby fruit bowl?

    a. They would grow toward the light.
    b. They would increase their rate of photosynthesis.
    c. They would ripen faster.
    d. They would exhibit secondary growth.
    e. They would take up chanting.

12. A plant devotes most of its energy budget to new growth during much of the summer. As the days begin to shorten (and the nights lengthen), it converts much of its effort to reproduction. This is an example of

    a. phototropism.
    b. photosynthesis.
    c. photoperiodism.
    d. phytophobia.
    e. philadelphia freedom.

BUT, WHAT'S IT ALL ABOUT?

Here's a question to help you pull together what you've learned so far using this text. If you don't remember how to attack this type of question, check out the example in Chapter 1.

Think back to the discussion of the 5 agents of microevolution (Chapter 17) and the descriptions of the 4 main categories of plants (Chapter 21). Describe what types of changes must have occurred in the ancestor of gymnosperms and angiosperms to drive the evolution of flowering plants. What subsequent changes have evolved to make angiosperms so successful?

What do I do now?
Remember the drill–decide what the question is asking you to do, collect your evidence from this chapter (and the others you've studied) and write!

CHAPTER 24

Form and Function in Flowering Plants

Basic Concepts

- Flowering plants vary in their anatomy and life cycles but share common cell types, transport systems, reproductive mechanisms, and growth patterns.
- Plant tissues are made up of three different kinds of cells—parenchyma, sclerenchyma, and collenchyma. These three tissue types comprise the four tissue types (dermal, vascular, meristematic, and ground) of all angiosperms.
- All angiosperms reproduce sexually through an alternation of generations.

Categorizing Flowering Plants

- Flowering plants can be classified by life span and by the number of embryonic leaves. Plants may complete their life cycles annually or biennially, or live for many years. Important food crops have a single embryonic leaf and are classified as monocotyledons. About 75 percent of all flowering plants are dicotyledons—having two embryonic leaves.

THINK ABOUT IT

1. Describe the two ways of classifying plants. Which method would the home gardener most likely use?

2. In what ways do monocots and dicots differ?

Plant Cells and Tissues

- Parenchyma cells comprise most of the mature plant's living tissues. Parenchyma cells are multipotent—capable of serving as the precursors to sclerenchyma and collenchyma cells as well as dedifferentiating into embryonic tissues.
- Sclerenchyma cells provide structural support for the plant; their walls are strengthened by lignin as well as cellulose. Sclerenchyma cells are dead when mature.
- Collenchyma cells share the structural properties of sclerenchyma cells and the growth potential of parenchyma cells.

- Herbaceous plants exhibit only primary growth—vertical growth from the tips of the roots and shoots. Woody plants, those that develop bark, exhibit both primary and secondary growth; secondary growth increases the plant's girth.
- Dermal tissue controls the plant's interactions with the outside world, regulating water and gas exchange, preventing infections, and secreting defense chemicals.
- Ground tissue, the most abundant tissue, is a prime site of photosynthesis.
- Vascular tissue, xylem and phloem, regulates the movement of water and food, respectively.
- Meristematic tissue is the source of the vascular, ground, and dermal tissues because it is the only tissue capable of dividing, growing, and differentiating.

THINK ABOUT IT

1. Label each of the following as belonging to parenchyma (P), sclerenchyma (S), or collenchyma (C).

   ____contains cellulose
   ____most abundant cell type
   ____thick-walled cells
   ____midway between the other two types

2. Which type of plant cell is dead when it provides its support functions?

3. Which type of plant cell is the basic starting plant cell, from which the other cell types differentiate?

4. Complete the following table describing the four main tissue types in plants and their basic function.

Tissue Type	Functions(s)
	Makes the outside covering of the plant
Vascular tissue	
	Makes up material responsible for growth
Ground tissue	

5. Which of the following have primary (P) and/or secondary (S) growth?

 ____trees ____herbs ____orchids

6. Where are apical meristems located, and what is their function?

7. If you accidentally chop off the top of your favorite plant, will it continue to grow?

8. Draw a root tip and label the zones of cell division, elongation, and differentiation.

9. What are the two zones responsible for secondary growth, and what do they produce?

10. What plant tissue type is wood composed of?

11. Why does a plant die if you remove a ring of its bark?

12. What are the specialized cell types that make up xylem?

13. Is water pushed or pulled through a plant? What causes water to flow from the roots to the leaves?

14. What are the main constituents of phloem sap? In which direction does phloem sap flow?

15. What structures in your body are analogous to xylem and phloem?

Sexual Reproduction in Flowering Plants

- Plants produce spores that grow into a distinct generation, producing gametophyte plants that look different from the parental plant. However, many of the sporophyte generation stay associated with the parental plants, so it is hard to distinguish these plants from their parents
- Gametophyte plants, the spore generation, produce haploid gametes (through meiosis) that fuse, producing a diploid zygote that will grow into a sporophyte plant.

- Flowers are the reproductive parts of the gametophyte generation, producing male (pollen grains) and female (eggs) gametes. Fusion of male and female gametes produces an embryo that becomes surrounded by fruit and a seed coat, both derived from the plant ovary.

THINK ABOUT IT

1. Fill out the following table of the two generations that alternate in an angiosperm life cycle.

Generation	Ploidy
Sporophyte	
Gametophyte	

2. What is meant by a double fertilization in angiosperms? What are the products of this double fertilization? Draw a flower undergoing double fertilization, and label the pollen tube and the ovule.

3. What are the differences between seed, fruit, and pollen?

4. What is the activation of the embryo and the beginning of growth called?

5. Which occurs first—the growth and establishment of a water source, or the growth and generation of a photosynthetic structure?

Cool Fact: Gimme Shelter

Sexual reproduction increases genetic diversity—plants whose eggs are fertilized by pollen of other plants, instead of self-pollinated, will produce seeds that carry slightly different genetic information than the parent plants. These small differences in genetic makeup may affect the plant's structure or functionality in favorable ways, giving this new plant a slight advantage over the parental species within its environment. Sexual reproduction is one of the contributing forces to evolution of species.

Because plants are sessile, they depend on pollinators to transport pollen from one plant to another. Plants use color, scent, and nectar rewards to attract pollinators, but certain species of philodendrons can offer pollinators heated accommodations. *Philodendron selloum* keeps its blossoms at a toasty 35°C, regardless of the surrounding environmental temperature. The warm blossoms release a scent that attracts the pollinator, the *Erioscelis emarginata* beetle. The beetles take advantage of the warm temperature overnight and leave the flower in the morning dusted with pollen that they can transfer to another philodendron the next night.

P. selloum pulls off this animal-like feat of warm "body parts" by selectively increasing the amount of oxygen that reaches the mitochondria within the blossoms to support heat production. The flowerets of *P. selloum* have stomata (the structures normally restricted to leaves) that allow gases in and heat and water vapor out. They have just enough stomata to permit an increased rate of cellular respiration without losing too much water vapor through evaporation. Increased cellular respiration increases the amount of energy dissipated as heat, making for toasty overnight accommodations for their pollinator.

Chapter 24

TESTING, TESTING

Make a web: clerenchyma, collenchyma, parenchyma, meristem, dermal, ground, xylem, phloem

Make a flowchart: Graph the stages of primary growth and secondary growth.

Vocabulary Review

Place the following terms in the appropriate columns of the table. Some terms may be used in more than one column.

cork | meristem | pollen
cork cambium | ovule | sieve elements
endosperm | phelloderm | sporophyte
gametophyte | phloem | tracheid | xylem

Growth	Transport	Reproduction

PRACTICE TEST

After you have finished studying this chapter, close your books, grab a pencil, and spend the next 15 to 20 minutes completing this practice test.

Compare and Contrast

For each of the following paired terms, write a sentence of comparison ("Both . . .") and a sentence of contrast ("However, . . . ").

pollen/endosperm

monocot/dicot

primary/secondary growth

cork cambium/vascular cambium

Matching

Match the following terms with their description. Each choice may be used once, more than once, or not at all.

a. develops from integument
b. helper cells in the phloem
c. develops from ovary
d. ground tissue storage area

____Fruit
____Seed coat
____Companion cells
____Pith

Short Answer

1. Explain the significance of this statement: "An ovary a day keeps the doctor away."

2. What is the primary function of meristem?

3. What is the direction of xylem flow within a plant?

4. What is a sporophyte?

Multiple Choice

Circle the letter that best answers the question.

1. Secondary xylem and secondary phloem form which of the following?

   a. epididymis
   b. bark
   c. gametophytes
   d. chloroplasts
   e. None of the above.

2. What type of cell forms the bulk of a herbaceous plant and carries out a variety of functions in that plant?

   a. collenchyma
   b. parenchyma
   c. meristems
   d. vascular bundles
   e. sclerenchyma

3. Which meristematic tissues are responsible for primary plant growth?

   a. vascular cambium
   b. cork cambium
   c. apical meristems
   d. all of the above
   e. a and b

4. What happens in the zone of differentiation?

   a. growth
   b. cell lengthening
   c. cell specialization
   d. phelloderm production
   e. None of the above.

5. What type of tissue makes up the bulk of a tree trunk?

   a. primary xylem
   b. primary phloem
   c. secondary xylem
   d. secondary phloem
   e cork

6. Which cells are dead at maturity?

   a. tracheids
   b. companion cells
   c. vessel elements
   d. all of the above
   e. a and c

7. What is the ploidy of endosperm?

   a. haploid
   b. diploid
   c. triploid
   d. aneuploid
   e. depends on the plant

8. A mature pollen grain consists of all of the following except:

   a. an outer coat of ground tissue.
   b. two sperm cells.
   c. two egg cells.
   d. one tube cell.
   e. all of the above are part of the pollen grain.

9. How do sperm reach the ovule?

   a. by entering through leaf stomata
   b. by entering through the root cortex
   c. by the pollen tube
   d. through ingestion by an animal pollinator
   e. by chloroplastic immigration

10. If the surrounding air was saturated with water (100% humidity), what would happen to xylem flow?

    a. It would move faster.
    b. It would reverse direction.
    c. It would slow or stop.
    d. Water content of air has nothing to do with xylem flow.
    e. Both a and b are correct.

11. After a 10-year absence, you return to the house where you grew up. Outside in the yard, you find that your old tree house is still in good shape. You decide to climb the ladder nailed into the side of the oak tree one last time to reminisce. You notice that 10 years of growth has caused the steps of the ladder to

    a. move upward (away from the ground).
    b. move downward (toward the ground).
    c. not change in height.
    d. win a basketball scholarship.
    e. None of the above.

12. Secondary growth of the tree will cause the steps to

    a. move downward (toward the ground).
    b. move upward (away from the ground).
    c. move closer to the center of the trunk.

d. move farther from the center of the trunk.

e. do the hokeypokey and turn themselves around (that's what it's all about!).

BUT, WHAT'S IT ALL ABOUT?

Here's a question to help you pull together what you've learned so far using this text. If you don't remember how to attack this type of question, check out the example in Chapter 1.

We've all had the experience of watching a well-watered plant die. Given what you've learned about transport in this chapter, explain why too much water can be a bad thing for your houseplants.

What do I do now?

Remember the drill – decide what the question is asking you to do, collect your evidence from this chapter (and the others you've studied) and write!

25

CHAPTER 25

Introduction to Animal Anatomy and Physiology: The Integumentary, Skeletal, and Muscular Systems

Basic Concepts

- Human bodies, like those of other mammals, are characterized by the presence of a central body cavity, an internal skeleton, and a stable body temperature.
- Humans have organs composed of specialized tissues. These organs function together, creating systems that carry out all the essential processes of the body—providing structural support, protecting the body against the environment, transporting nutrients in and wastes out, interacting with the environment, and maintaining a stable internal environment.
- Covering the body and maintaining body temperature is the function of the integumentary system, which is comprised of the skin, and its associated structures and glands.

Human Characteristics

- Multicellular animals, like humans, require systems to deliver nutrients and remove wastes from the cells that are not in direct contact with the environment.
- These systems, composed of organs, reside within a central body cavity (coelom). The shape of the body is maintained by an internal skeleton, which also protects the organs. As the organ systems carry out their metabolic functions, heat is generated. Thus, humans are able to maintain a stable internal temperature.
- The field of biology that describes the structures comprising a body is called anatomy; physiology examines the functions of these structures.

THINK ABOUT IT

1. Label each of the following as belonging to the study of anatomy (A) or physiology (P).

   ____The shape and placement of your left little toe.
   ____The conversion of doughnuts into energy sources.

____The transfer of messages from the brain to the muscles.
____A comparison of eye position in monkeys, cows, and salamanders.

2. Which body cavity dorsal (D) or ventral (V) contains:

   ___the food from your last meal?
   ___the systems you use to understand these words?

3. a. Is a cooking pot ectothermic or endothermic? What about the stove it sits on?

   b. Are reptiles endothermic or ectothermic? What does this mean about their body temperature?

4. Put the following in order of simplest to most complex: organs; tissues; organ systems; cells; molecules.

5. Of the terms listed in question 4, which refers to the smallest unit of life? Which term describes the heart? Which term describes all the blood vessels in the body?

Tissues, Organs, and Organ Systems

- Four different types of tissues exist in the human body: epithelial, connective, muscle, and nervous.
- Organs are collections of tissues that cooperate to perform a single function. Multiple types of tissues may be found in a single organ, in order to perform different steps of the metabolic process carried out in the organ.
- Organs work in concert to carry out bodily functions. There are 11 organ systems in the human body: integumentary, skeletal, muscular, nervous, endocrine, lymphatic, cardiovascular, respiratory, digestive, urinary, and reproductive.

THINK ABOUT IT

1. Complete the table to indicate the functions of some of the organ systems:

Organ System	Role
Lymphatic	
Cardiovascular	
Urinary	
Integumentary	

2. Which type of tissue:

   a. forms the internal fluid transport system?

   b. transmits messages along a thin cell extension?

   c. makes up the biceps?

   d. forms the outer boundary of the body?

   e. contains the stored excess energy from your chocolate excesses?

   f. gives distinctive shapes to your ears or your nose?

3. Complete the following table describing the four animal tissue types, their basic functions in the animal body, and a specific example of each.

Tissue	Function	Example

4. Why do epithelial cells contain keratin?

5. Mark the following muscles as striated (ST) or smooth (SM):

_____heart muscle ____biceps of arm _____muscles around bladder

6. Draw a typical neuron and label its structures.

The Integumentary System

- Our skin is the largest organ of our bodies. It is made of two layers, the dermis and the epidermis, and serves as the first line of defense, protecting the internal organs from the environment. Hair, sweat and sebaceous glands, and nails are skin-associated structures that help us regulate our body temperature (hair and the glands) and provide protection.
- The skeletal system counters the effects of gravity. Bone tissue is mostly connective tissue strengthened by large deposits of calcium compounds. Within the bone exists a loose connective tissue, marrow, that makes blood tissue. Bones connect at the joints, highly mobile structures that hold bones together without the bones touching, so that the body has flexibility.
- Muscle produces movement by contraction and relaxation. During a contraction, muscle fibers (collections of muscle-tissue cells) slide past each other,

bringing the attached bones closer together. When the muscle relaxes, the fibers slide apart, separating the bones.

THINK ABOUT IT

1. What are two types of skin glands and their secretions?

   a.

   b.

2. Your cat raises its hackles (a ridge of fur along the spine) at the sight of another male cat. What muscles are involved in this response?

3. A new deodorant is designed to inhibit body odor. Should it interact with apocrine or merocrine sweat glands?

4. What types of tissues are transplanted in the following operations?

   a. bone-marrow transplant

   b. blood transfusion

   c. skin graft

5. a. Which three cell types are found in bone?

   b. For each of the following ages, give the most common bone cell type:

   ____8 months
   ____28 years
   ____48 years
   ____68 years

   c. Which bone cell type likely contributes to osteoporosis?

6. Why do many more injuries occur at joints than in long bones?

7. Why do you think that the human knee joint has three padding systems (fluid, cartilage, and fat)?

8. You can approximate the action of a muscle sarcomere by imagining that the fingers of your right hand are the thick (myosin) filaments and the fingers of your left hand are the thin (actin) filaments. Try this: (1) With your palms facing you, interlace just the fingertips of your hands. (2) Slide your fingers toward each other.

   a. What happens to the overall length of the "sarcomere" (thumb tip to thumb tip)?

   b. How could this action allow a muscle to lift objects?

   c. How could this action pump blood?

9. a. Muscles contract by the movement of actin and myosin. Place the following steps in the correct order:

   ____Pull actin filament to middle of the sarcomere.
   ____Myosin heads attach to actin filament.
   ____Myosin heads detach from actin filament.

   b. What is the energy source for muscle contraction?

Cool Fact: Burning Issue

Ten thousand people die of burns in the United States every year. Surviving burns depends on how seriously the skin has been damaged. First- and second-degree burns damage the epidermis and leave the accessory structures unaffected, but third-degree burns damage the dermis layer, destroying accessory structures, nerves, and blood vessels. Third-degree burns represent the greatest threat, because not only is the tissue unable to repair itself, but it also leaves the body open to infection. Fluids are readily lost through the burned area, making the burned area damp—a good habitat for bacterial growth. Losing fluids also alters the patient's electrolyte balance and makes it harder to maintain a stable body temperature; more metabolic effort must be expended to counter the cooling effect of fluid evaporation.

Third-degree burns must be treated surgically, which traditionally meant grafting skin from one part of the body to the burned area. This need to graft tissue meant that patients with large burns often had the poorest prognosis for recovery. Fortunately, advances in cell-culturing techniques have made it possible to grow large sheets of epidermal cells from small samples. Using cultured cells in skin grafts, as well as improved methods of controlling infection and maintaining fluid and electrolyte imbalance, survival probabilities for seriously burned people have increased dramatically. Continued improvement in culturing techniques, and development of "artificial" skin to restore the skin's integrity while waiting for a graft, should have an additional impact on patient survival.

TESTING, TESTING

Create a web: muscle, epithelial, connective, nervous, integumentary system, skeletal system, muscular system

Make a flowchart: Place these terms in order of size from smallest to largest, and explain how each level gives rise to the next: cells, organ systems, tissues, organs Santa Claus.

Vocabulary Review

Place the following terms in the appropriate columns of the table. Some terms may be used in more than one column.

cardiovascular endocrine lymphatic nervous skeletal
digestive integument muscle respiratory urinary

Both Support and Movement	Coordination, Regulation, Defense	Transport and Exchange with the Environment

PRACTICE TEST

After you have finished studying this chapter, close your books, grab a pencil, and spend the next 15 to 20 minutes completing this practice test.

Compare and Contrast

For each of the following pairs of terms, write one sentence describing their similarities and one sentence explaining their major differences. In writing about similarities, start the sentence with "Both …"; for differences, start with "However …"

ectotherm/endotherm

exocrine/endocrine glands

basement membrane/ground substance

organelle/organ

Matching

Match the following terms with their descriptions by writing the appropriate letters from column two in the blanks.

a. cardiac muscle
b. stratified squamous epithelium
c. connective tissue
d. message transfer and storage
e. forms linings and coverings

___Bone
___Nervous
___Heart
___Epithelial
___Skin

Short Answer

1. How are connective tissues defined?

2. Explain the functions of the three types of muscle tissue.

3. What is the function of the exocrine glands? Give an example.

4. Explain the difference in function between red and yellow bone marrow.

Multiple Choice

1. There are ______main tissue types in animals.

   a. 1
   b. 2
   c. 3
   d. 4
   e. 5

2. Blood is a type of

   a. elastic tissue.
   b. epithelium.
   c. muscle.

d. connective tissue.
e. nervous tissue.

3. What mineral helps to harden bone?

a. magnesium
b. TUMs
c. calcium
d. zinc
e. iron

4. Sweat is produced by

a. apocrine glands.
b. sebaceous glands.
c. hair follicles.
d. keratin.
e. a and b

5. Which of the following is not a type of connective tissue?

a. blood
b. bone
c. ligaments
d. skin
e. tendons

6. Which of the following is NOT a characteristic of mammals?

a. internal body cavity
b. temperature regulation
c. internal skeleton
d. single-celled organisms
e. hair

7. Connective tissue

a. includes muscle cells.
b. is the most diverse of the tissue types.
c. is found only among plants.
d. is classified as squamous, cuboidal, and stratified.
e. All of the above.

8. The function of the lymph system is to

a. produce hormones.
b. produce skin cells.
c. protect the body from invasion.
d. transmit messages.
e. make blood cells.

9. What do the following systems have in common? The cardiovascular, respiratory, digestive, and urinary systems all:

   a. are comprised solely of connective tissue.
   b. have no organs.
   c. have structures within the skull.
   d. transport and exchange materials with the environment.
   e. All of the above.

10. Every organ system

   a. is made up of several different types of tissues.
   b. can survive independently outside the body.
   c. is classified as a -cyte, -blast, or –clast.
   d. has cells suspended in liquid material.
   e. has pipes and pedals.

11. Your friend Serafina had a minor accident last week while rock climbing. Since then, she feels a sharp pain whenever she tries to use her left index finger. She thinks she may have broken a bone. You suggest that she consult a doctor who specializes in

   a. muscle tissue
   b. epithelial tissue
   c. connective tissue
   d. nervous tissue
   e. facial tissue

12. Scleroderma ("hard skin") is a disease that affects epithelial tissue. Aside from the skin, what other structures would you predict would be affected?

   a. lining of the stomach
   b. lining of the lungs
   c. heart muscle
   d. both a and b
   e. both b and c

BUT, WHAT'S IT ALL ABOUT?

Here's a question to help you pull together what you've learned so far using this text. If you don't remember how to attack this type of question, check out the example in Chapter 1.

In the previous unit, we learned about the basic structure and function of plant tissues. Can we identify similar types of tissues in animals? Do they have similar functions?

What do I do now?

Remember the drill—decide what the question is asking you to do, collect your evidence from this chapter (and the others you've studied) and write!

26

CHAPTER 26

Control and Defense: The Nervous and Endocrine Systems

Basic Concepts

- The nervous system processes information received from both inside and outside the body through a network of sensory neurons. Information is processed either by the central nervous system (CNS) or directly by the peripheral nervous system (PNS). The autonomic and somatic divisions of the PNS carry information directing a response to the appropriate muscles and organs.
- The endocrine system produces the hormones, the chemical messengers responsible for maintaining internal homeostasis in the face of a constantly changing external environment.

The Nervous System

- The nervous system is composed of the brain and spinal column (CNS) and the neurons that transduce information to and from the responding organs (PNS). Some responses to nerve impulses are under voluntary control (somatic), and some are nonvoluntary (autonomic).
- Primary cells of the nervous system are neurons; sensory neurons deliver information from the sensory organs, motor neurons deliver information to muscles, and interneurons carry information between the two. The protective neuroglia cells surround neurons in the CNS.
- Information travels in the nervous system as a change in the distribution of charge across the membrane; this change in membrane potential is called an action potential because it travels unidirectionally from a neuron cell body to the end of the axon.
- Neurons do not touch, so action potentials must use neurotransmitter molecules to travel across the small space (synapse) separating one axon from the cell body of another neuron.
- The spinal cord serves as a conduit for messages between the PNS and the brain. Some incoming messages to the spinal cord elicit immediate, efferent responses to the muscles—these are the reflexes, the simplest response of the autonomic nervous system.
- Involuntary responses may either conserve energy (parasympathetic response) or stimulate the demand for energy (sympathetic response). Both divisions of

the autonomic nervous system innervate organs, so organ function represents a balance between the two messages.

THINK ABOUT IT

1. We can think about the parts of the nervous system as branches on a tree. Use the terms described in your textbook in Section 26.1 to complete the tree on the left.

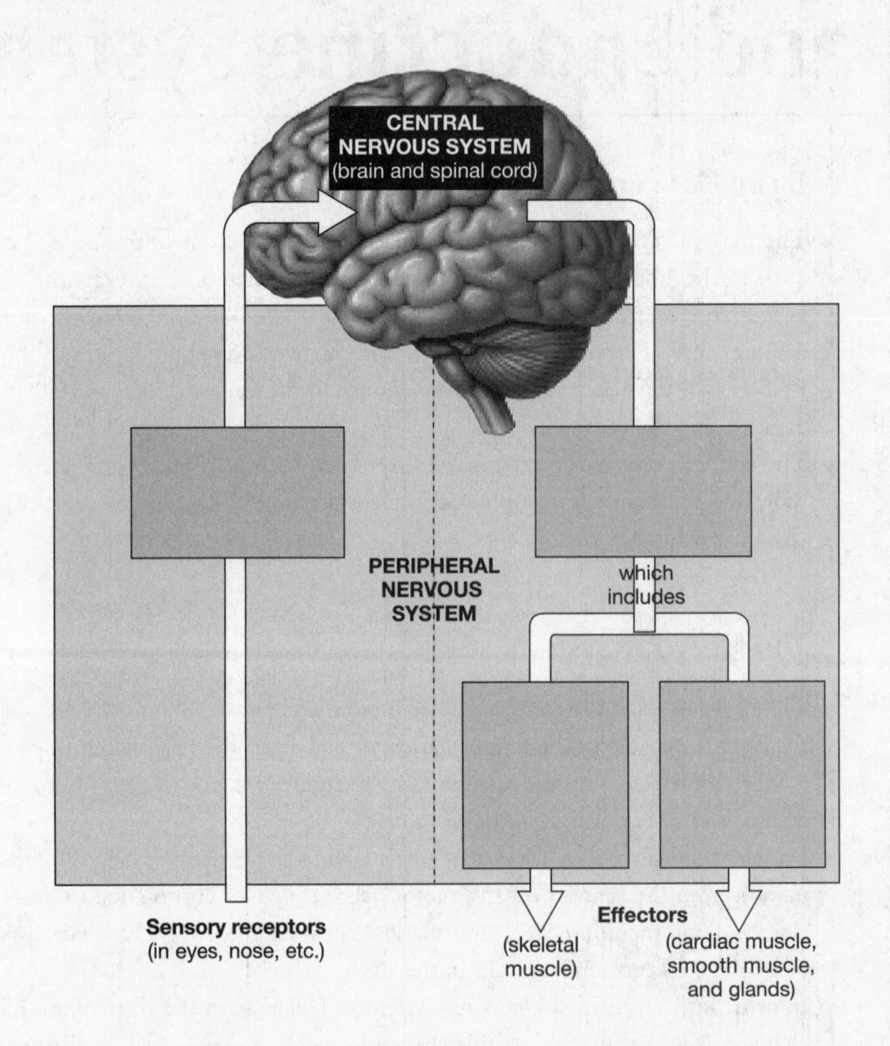

2. Match the terms in the table below to the descriptions of some of their functions: peripheral nervous system, central nervous system, afferent nerves, efferent nerves.

Function	Term
Interpreting these words as you read them	

Function	Term
Carrying the "pain" message from your burned finger to your brain	
Carrying messages to and from the extremities	
The nerves that cause you to pull your hand away from a hot stove	

3. What is the role of neuroglia cells?

4. Which division of the PNS helps sense stimuli? Which division helps respond to stimuli?

5. Complete the following table to indicate the differences between the three types of neurons found in the nervous system.

Neuron Type	Role
Sensory	
	Carry information from the CNS to tissues
Interneurons	

6. a. What is the purpose of myelin?

   b. What do myelin and the blood-brain barrier have in common?

7. As you turn your head, you smell a rose. You reach out to pick it up. Explain the roles of motor neurons, sensory neurons, and interneurons in this scenario.

8. Draw the distribution of negative and positive charge inside and outside a neuron. What is this charge distribution called? Which ions are responsible for the charge distribution?

9. Describe the ion flow associated with an action potential. How does an action potential travel down an axon?

10. Complete the following table describing the role of different components of the synapse in signal transmission.

Component	Role in Transmission between Neurons
Presynaptic neuron	
Synaptic cleft	
Postsynaptic neuron	

11. Number the following structures in the correct order for signal transmission:

    ___ neurotransmitter ___ postsynaptic neuron

    ___ presynaptic neuron ___ synaptic cleft

12. What do neurotransmitters do? Do all neurotransmitters send the same message?

13. For the following scenarios, label each as being most similar to an action potential or synaptic transmission.

    a. a row of dominoes that fall sequentially, finally ringing a bell at the end of the line

    b. a Pony Express rider tossing the mail pouch across a stream to the courier waiting on the other side

14. Why do you pull your hand back so quickly after being stuck by a pin? Describe the parts of the reflex arc that are required for that action to occur.

15. a. You sit down in your beanbag chair to read your new novel and sip herbal tea. Which division of the autonomic nervous system is most active?

    b. You suddenly realize that you've left the stove on, and smoke is billowing from the kitchen. Which branch of the autonomic nervous system is activated as you race for the fire extinguisher?

16. Why do most organs receive instructions from both the rest-and-repose (parasympathetic) and fight-or-flight (sympathetic) nervous systems?

17. Which part of the brain do you use for the following activities?

Studying biology	______________
Breathing	______________
Listening to a biology lecture	______________
Watching TV	______________

18. Why is it that a person can survive in a vegetative state after massive trauma to the cerebrum as long as the brain stem (thalamus, midbrain, pons, medulla oblongata) is intact?

19. Use the following terms to fill in the numbered boxes: sympathetic, somatic, spinal cord, PNS, brain, afferent.

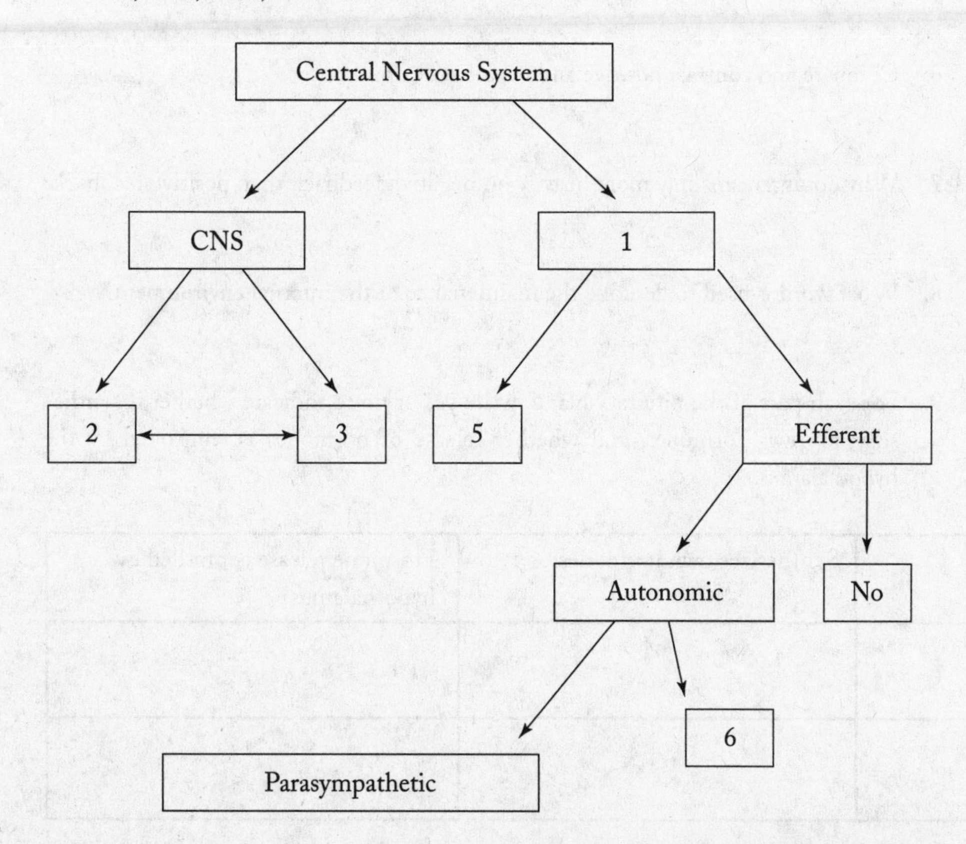

The Endocrine System

- Organs that synthesize chemical messengers exert their effect on some other tissue.
- Hormones may be either amino-acid-based or lipid-based. Amino-acid hormones must be detected by receptors on the surface of the target cell, whereas lipid-based hormones can diffuse through cell membranes.

THINK ABOUT IT

1. Which type of gland secretes hormones, and what is unique about these glands? (Hint: think back to Chapter 25.)

2. Which type of hormone is derived from cholesterol? Can you see why a certain amount of cholesterol is necessary?

3. How do hormones travel throughout the body to reach their site of action?

4. Draw a cell. Using different colors, indicate how amino acid-based and steroid-based hormones each get their message into the cell.

5. What effects do hormones have on cells?

6. Compare and contrast positive and negative feedback.

7. Why do organisms rely more heavily on negative feedback than positive feedback?

8. What word is used to describe the maintenance of the internal environment?

9. For each part of the pituitary gland, write yes or no to indicate whether it synthesizes its own hormones, and whether release of hormones is controlled by the hypothalamus.

Pituitary:	Synthesizes own hormones?	Hormone release controlled by hypothalamus?
Posterior		
Anterior		

10. Which anterior pituitary hormone is responsible for ovulation in human females?

11. Growth hormone may be made by genetically engineered bacteria for pharmaceutical production. Where is growth hormone made in the human body?

Cool Fact: Phytoestrogens

Plants also have an endocrine system, although they lack a nervous system. The endocrine molecules of the plant world are the flavenoids, a class of compounds secreted by plants to discourage herbivore activity, protect the plant from pathogens, and, in leguminous plants, to attract nitrogen-fixing bacteria into symbiosis. Flavenoids allow the plant to respond to environmental changes.

Flavenoids also act as signaling molecules when ingested by animals. The flavenoids genistein and luteolin can bind to the vertebrate estrogen receptors and elicit an estrogen response in sensitive tissues. The observation that vertebrate receptors are responsive to these plant estrogens—hence the name phytoestrogens—suggests that endocrine signaling systems must have developed before the evolution of animals from early protists.

This responsiveness of animal tissues to phytoestrogens is a mixed blessing. Consumption of foods rich in phytoestrogens, such as soy products, is believed to reduce the risk of breast cancer and moderate the "estrogen-withdrawal" symptoms of menopause. But high levels of dietary phytoestrogens cause infertility in sheep and mice. No direct evidence exists that phytoestrogens cause beneficial effects in humans, only that phytoestrogen consumption correlates with decreased risk of disease. Until the benefits can be shown to be a direct result of eating phytoestrogens, people would be wise to proceed cautiously with any dietary modifications.

Chapter 26

TESTING, TESTING

Create a web: nervous, endocrine, communication, control, cells, receptor, hormone, neurotransmitter

Vocabulary Review

Place the following terms in the appropriate columns of the table. Some terms may be used in more than one column.

axon	hypothalamus	neurotransmitter	prolactin
gland	negative feedback	pituitary	steroid
homeostasis	neuroglia	pons	synapse

Nervous System	Endocrine System

PRACTICE TEST

After you have finished studying this chapter, close your books, grab a pencil, and spend the next 15 to 20 minutes completing this practice test.

Compare and Contrast

For each of the following pairs of terms, write one sentence describing their similarities and one sentence explaining their major differences. In writing about similarities, start the sentence with "Both …"; for differences, start with "However …"

nerve/neuron

steroid hormone/amino acid hormones

pituitary/hypothalamus

axon/dendrite

Matching

Match the following terms with their descriptions by writing the appropriate letters from column two in the blanks.

a. Autonomic Nervous System
b. Hormone
c. action potential
d. synapse
e. pituitary

_____ Message carried by an axon
_____ Control center of the endocrine system
_____ Site of message transfer in the nervous system
_____ Regulation of smooth muscle
_____ Message received via the bloodstream

Short Answer

1. Explain the difference in the communication methods of the nervous and endocrine systems.

2. Explain the relationship of rods and cones.

3. How is homeostasis maintained in the body?

4. What does the length of the eyeball have to do with the ability to focus?

Multiple Choice

Circle the letter that best answers the question.

1. Which of the following represents the correct direction of signal movement through a neuron?

   a. axon → cell body → dendrite
   b. dendrite → cell body → axon
   c. cell body → axon → dendrite
   d. axon → dendrite → cell body
   e. dendrite → axon → cell body

2. What common molecule forms the core of steroid hormones like testosterone?

   a. collagen
   b. cellulose
   c. cholesterol
   d. calcium
   e. cheese curls

3. The brain and spinal cord are part of the

   a. peripheral nervous system.
   b. autonomic nervous system.
   c. parasympathetic nervous system.
   d. sympathetic nervous system.
   e. central nervous system.

4. What are neurotransmitters used for?

   a. transmission of an action potential along an axon
   b. transmission of an action potential between a dendrite and an axon
   c. action potential transmission from an axon to a cell body
   d. action potential transmission between two neurons at a synapse
   e. to sense the environment

5. If you sustained an injury to your cerebellum, which of the following might you have difficulty with?

   a. breathing
   b. keeping balance
   c. studying
   d. sensing information from the external environment
   e. There would be no consequences of an injury to the cerebellum.

6. What controls the anterior pituitary?

   a. posterior pituitary
   b. ADH
   c. thalamus
   d. hypothalamus
   e. oxytocin

7. The brain knows that light has stimulated the rods and cones of the eye when

   a. the rods and cones change color.
   b. only the cones increase the rate of signalling.
   c. only the rods increase the rate of signalling.
   d. the rods and cones stop signalling.

8. The complex that allows the transfer of an action potential from a sensory neuron to a motor neuron or interneuron is called a

   a. neuroglia.
   b. neuropathy.
   c. synapse.
   d. complexion.
   e. cortex.

9. All of these are part of a reflex arc except:

   a. sensor receptor.
   b. effector (muscle or gland).
   c. spinal cord.
   d. cerebral cortex.
   e. All of the above are required for a functional reflex.

10. Which of the following is the brain structure that allows you to perceive and answer this question?

    a. cerebellum
    b. pons
    c. hypothalamus
    d. cerebrum
    e. thalamus

11. Which of the following is a true statement about the endocrine system?

    a. Most hormones are made from carbohydrates.
    b. Most hormones act only where they are synthesized.
    c. Every hormone has a specific target that it affects.
    d. Neurotransmitters are DNA-based hormones.
    e. Endocrine glands are composed of mesothelial tissues.

12. Which of the following is the best example of negative feedback?

    a. Drinking salty seawater makes you thirsty, so you drink more seawater.
    b. Uterine contractions during childbirth send signals to the brain that increase the rate and intensity of contractions.
    c. Eating one potato chip gives you cravings for more.
    d. The urge to skydive is removed by jumping out of a plane.
    e. Heavy rains cause flooding.

13. Which of the following would be a homeostatic mechanism?

    a. receptors that make you feel more tired as you increase the number of hours you sleep each night
    b. receptors that cause an increase of blood flow to an area with an open wound
    c. receptors that respond to a drop of blood iron levels and send a message to the brain that causes you to crave spinach (an iron-rich food)
    d. receptors that respond to the color of spring flowers
    e. None of the above.

BUT, WHAT'S IT ALL ABOUT?

Here's a question to help you pull together what you've learned so far using this text. If you don't remember how to attack this type of question, check out the example in Chapter 1.

The visual system has 3 component tasks, which simplify conceptually to "convert light signals into information we can use." Being able to use that information—climb a stair, avoid a snake, pick an apple—is what matters to us. Occasionally, people who have lost their vision when very young regain it after surgery as adults. Despite being able to see, however, these people often find it harder to make sense of their world – why?

What do I do now?
Remember the drill—decide what the question is asking you to do, collect your evidence from this chapter (and the others you've studied) and write!

27

CHAPTER 27

Immune Systems

Basic Concepts

The immune system employs nonspecific and specific defenses to protect the body from foreign organisms and toxins. Nonspecific defenses block or attack invaders indiscriminately, whereas specific defenses target invaders and create a "memory" within the immune system to combat future attacks.

The Immune System

- Nonspecific defenses of the immune system do not employ invader-specific mechanisms. Physical barriers to penetration by microorganisms, phagocytic cells, and antiviral compounds define most of the nonspecific "arsenal."
- Specific defenses employ T and B cells, and their products, to eliminate invading organisms. T and B cells recognize features of the invader in much the same way that receptors recognize their ligands. When an invader binds, a response is triggered, which could include the release of cytokines or the production of antibodies.
- Immunity can be acquired passively or actively. Passive immunity comes from microorganism compounds that are made outside of the body and introduced into the body for the short-term goal of eliminating the invader (antigen). Active immunity is long-lasting protection gained after a previous exposure to the invader, either because of previous illness or vaccination.
- Actively acquired immunity comprises the action of cell-mediated and antibody-mediated processes. Cell-mediated processes involve activated T cells that produce antigen-fighting chemicals or become phagocytotic. Antibody-mediated immunity relies on B cells to produce specific antibodies that bind to the antigen and drive its removal from the body. Both immune processes can build a memory, so that subsequent exposure to the antigen reactivates the immune response.
- Occasionally the body mistakes its own tissues for foreign antigens and elicits an immune response against itself. Allergic reactions are one form of autoimmune response; rheumatoid arthritis, lupus, and diabetes are others. Autoimmune responses can be deadly.
- In AIDS (acquired immunodeficiency disease syndrome) the human immune system has been attacked by the invading virus, thus eliminating the system that could defeat the virus.

THINK ABOUT IT

1. What is the difference between specific and nonspecific defenses? How are they similar?

2. List the mechanisms of the nonspecific responses for each of the following categories.

Physical Barriers	Cellular	Molecular	Physiological Responses

3. How is skin well suited to prevent entry of foreign substances into the body?

4. What is a phagocyte?

5. What type of cell is responsible for detecting a virally infected cell with viral proteins on its surface?

6. Which of the following are examples of passive (P) immunity and which are examples of active (A) immunity?

   — an infant drinking its mother's milk
   — vaccination with an inactivated poliovirus
   — injection of anti-rabies antibodies after a dog bite

7. Complete the following table highlighting some of the important distinctions between antibody- and cell-mediated immunity.

	Cell Type(s)	Antibody production?	Direct contact with target cell?
Antibody-mediated			
Cell-mediated			

8. a. Why don't you get chicken pox twice in a lifetime?

   b. Why do you get so many colds during your lifetime?

9. a. Which cells produce antibodies?

   b. Which cells are responsible for the memory of the immune system?

10. Circle the cell types that are involved in cell-mediated immunity:

helper T cells	plasma cells	natural killer cells
cytotoxic T cells	memory B cells	

11. Why do allergies resemble an immune-system disorder?

12. What actually kills a patient with AIDS?

13. Why is the CD4 count so important for an AIDS patient?

14. What causes AIDS?

15. How do protease inhibitors work?

16. **Make a flowchart:** from first exposure to the antigen to the last cell destroyed by a killer T-cell, follow the path of an infection.

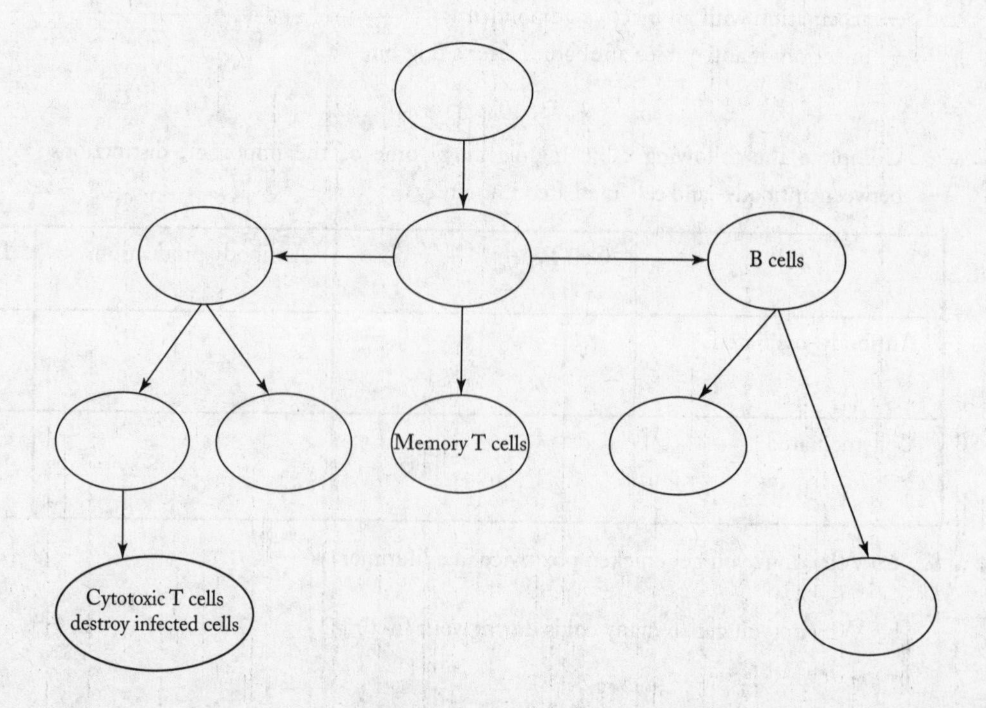

Cool Fact: Naked DNA, Oh My!

Childhood immunization programs have made many viral and bacterial diseases rare occurrences in the developed world, but this protection does come with some risk. Occasionally, the weakened virus or bacteria used to elicit an immune response in a vaccine regains its strength when injected and produces a case of the full-blown disease. Rare as these episodes are, they remain a real enough risk to dissuade parents from vaccinating their children. Imagine if it were possible to produce a robust and long-lasting B- and T-cell mediated immunity without needing to use the disease-causing organism.

Such a method was discovered, by accident, in 1993 when a group of researchers experimenting with gene therapy discovered that DNA plasmids containing genes for influenza proteins produced both a B-cell and T-cell response when injected into the muscles of mice. Apparently, the genes carried by the plasmid were transcribed and translated into active proteins by the muscle cells and elicited a strong immune response. Although a setback for gene therapy, the ability of "naked DNA" to serve as vaccines promises to revolutionize preventive medicine. Since the early 1990s, DNA vaccines have been developed against hepatitis B and C, HIV, tuberculosis, dengue fever, and malaria. The malaria vaccine completed phase I human trials (first of 3 phases) in 1999; the other vaccines mentioned are still being developed. The technology may also help scientists develop vaccines against some types of cancers. Perhaps, thanks to DNA vaccines, the heartbreak of many infectious (and non-infectious) diseases will become distant memories.

Chapter 27

TESTING, TESTING

Create a web: macrophage, cytokines, skin, helper T cells, cell-mediated, allergens, antibody-producing, autoimmune

Vocabulary

antibody	B cell	histamine	inflammation	second line	T cell
antigen	first line	immunity	macrophage	sweat glands	vaccine
	helper T cell				

Specific	Non-specific

PRACTICE TEST

After you have finished studying this chapter, close your books, grab a pencil, and spend the next 15 to 20 minutes completing this practice test.

Compare and Contrast

For each of the following paired terms, write a sentence of comparison ("Both . . . ") and a sentence of contrast ("However, . . . ").

specific defense/non-specific defense

B cell/T cell

histamine/cytokine

dendritic cells/macrophages

plasma cells/B cells

Matching

Match the following terms with their description. Each choice may be used once, more than once, or not at all.

a. interferon
b. allergen
c. antigen
d. dendritic cells
e. B cell
f. helper T cell

— Causes the activation of natural killer cells and macrophages
— Causes the release of histamine
— Stimulates both B cells and T cells
— Stimulates the formation of plasma cells
— Present antigens to other cells
— Generates the formation of antibodies

Short Answer

1. What are autoimmune disorders?

2. What is a vaccine?

3. What is the inflammatory response?

4. Explain the difference between polio and AIDS.

Multiple Choice

1. Which of the following is NOT a molecule used by the non-specific defenses?

   a. lysozyme
   b. allergen
   c. complement
   d. interferon
   e. histamine

2. How does a natural killer cell recognize cancer cells?

   a. by their location
   b. by their accelerated growth rate
   c. by their altered cell surface proteins
   d. by their mutated DNA
   e. None of the above.

3. What is the major benefit of the specific defense system?

   a. Specific defense systems act as barriers to foreign invaders.
   b. Specific defenses provide a quicker response than non-specific defenses.
   c. Specific responses are generated no matter what the situation is.
   d. Specific defenses can produce immunity.
   e. All immune responses are specific.

4. Which of the following cells engulf and digest other cells from your body?

   a. helper T cells
   b. suppressor T cells
   c. B cells
   d. macrophages
   e. Schwann cells

5. What do rheumatoid arthritis and multiple sclerosis have in common?

   a. They are both found only in individuals of Asian descent.
   b. They are both diseases of the nervous system.
   c. They are both forms of small-cell cancer.

d. They are both the result of malfunctions of the immune system.
e. They are both caused by bacterial infections.

6. What system is attacked by the human immunodeficiency virus?

a. nervous
b. reproductive
c. nonspecific immune
d. specific immune
e. the blood-cell-generating bone marrow

7. Serafina is fighting off an infection. She has managed to drag herself over to visit you despite the fever and other symptoms she is suffering. She vows to have a doctor cut out her swollen, painful, lymph glands and thereby remove the source of her illness. You explain that it would be a bad idea, because

a. The glands are filled with immune-system cells fighting the infection.
b. Those little pathogens get really mad when you destroy their dens.
c. Lymph glands are found only in the central portion of the cerebral cortex in the brain.
d. It is better to wait until they are fully ripe to pick them.
e. The doctor will charge too much, and you can do it for free with your butter knife.

8. Activation of which of the following is required for the production of antibodies?

a. B lymphocytes
b. Helper T cells
c. plasma cells
d. antigens
e. All of these play a role in generating antibodies.

9. Which of the following is NOT a symptom of an inflammation?

a. redness
b. chills
c. pain
d. swelling
e. heat

10. Which is the proper sequence of function for T cells?

a. Suppressor T → Helper T → Cytotoxic T
b. Helper T → Suppressor T → Cytotoxic T
c. Cytotoxic T → Suppressor T → Helper T
d. Helper T → Cytotoxic T → Suppressor T
e. Suppressor T → Cytotoxic T → Helper T

11. Passive immunity

    a. can be passed from mother to child.
    b. can be generated only for bacterial infections.
    c. can be administered as an injection.
    d. can be found only in cows (*vacca*).
    e. both a and c

12. What actually stimulates helper T cells to begin the specific immune response?

    a. allergens
    b. cytokines
    c. antibodies
    d. histamine
    e. lymph

BUT, WHAT'S IT ALL ABOUT?

Here's a question to help you pull together what you've learned so far using this text. If you don't remember how to attack this type of question, check out the example in Chapter 1.

In recent times, we have seen an increase in the number of people suffering from asthma and allergies, especially young children. Some people have suggested that this has happened because the immune system was not adequately stimulated early in life, that is, toddlers aren't seeing enough "dirt" because of our air-conditioned homes, antibacterial cleaners, and limited outside play. Given what you know about specific immunity, is there any support for this hypothesis?

What do I do now?

Remember the drill—decide what the question is asking you to do, collect your evidence from this chapter (and the others you've studied) and write!

28

CHAPTER 28

Transport, Nutrition, and Exchange: Blood, Breath, Digestion, and Elimination

Basic Concepts

- The cardiovascular system transports gases, nutrients, and waste products throughout the body using muscular contractions of the heart.
- The digestive system extracts usable material from ingested food and disposes of the waste.
- The respiratory system primarily functions to supply oxygen to the body's tissues and remove carbon dioxide, but it also provides immune surveillance, regulates blood pressure and pH, and provides structures for sound generation.

The Cardiovascular System and Body Transport

- Blood consists of cells suspended in a protein-rich plasma. The red blood cells transport oxygen and carbon dioxide, and the white blood cells perform immune functions. The plasma transports hormones and fatty acids bound to proteins, and carries nutrients and wastes dissolved in solution.
- Blood vessels have an inner layer of epithelial cells, a middle layer of smooth muscle, and an outer layer of connective tissue. Vessels that carry blood away from the heart are arteries and arterioles; veins return blood to the heart. Valves maintain unidirectional flow in the veins.
- The heart sits at the center of two circulation systems—pulmonary and systemic. In the pulmonary circulation, blood is pumped into the lungs by the right ventricle and receives oxygenated blood from the lungs into the left atrium. In the systemic system, blood from the left atrium flows into the left ventricle and is pumped out to the body. Blood returning from the body enters the right atrium through the vena cavae.
- Like any tissue, the heart is surrounded by coronary arteries that supply this muscle with oxygen and nutrients. Blockages within those arteries starve the heart tissue of essential nutrients and cause tissue death; this is a myocardial infarction—a heart attack.
- Nutrients and gases are delivered to the tissues in the capillary beds. Pressure resulting from the heart's contractions push gases and small molecules through the capillary walls and into the interstitial fluids. Osmotic pressure works to move water and small molecules (O_2, CO_2, urea) back into the veins.

THINK ABOUT IT

1. Assign these terms to the appropriate column (Plasma or Cells):

   red blood cells   fibrinogen   white blood cells
   albumin   platelets   globulins

Plasma	Cells

2. What is the difference between plasma and serum?

3. Why do the largest animals on the planet have closed circulatory systems?

4. Draw a cross section of an artery and label the three layers.

5. How does the cross section of a capillary compare to the cross section of an artery? How does it compare to the cross section of a vein?

6. Why does the human heart contain four chambers? List these chambers and describe the function of each. What is the order of blood flow through the heart?

7. Which circulatory system picks up oxygen from the lungs and which circulatory system drops off oxygen at the tissues of the body?

8. Which do you think is the strongest chamber of the heart? Why?

9. What vessels are blocked in a heart attack?

10. What is the correct order for these terms? heart, venules, arteries, capillaries, veins, arterioles

11. Where does exchange of gases between blood and cells of the body take place? What other system assists in returning blood back toward the heart?

12. Why is osmosis important in the exchange of nutrients between blood and the cells of the body?

13. Place the following terms in the appropriate column (Arterioles or Venules): CO_2, glucose, wastes, O_2, red blood cells, sodium, nitrogenous wastes

Arterioles	Venules

14. Label the figure with the following terms: right ventricle, left ventricle, pulmonary arteries, pulmonary veins, aorta, left atrium, right atrium

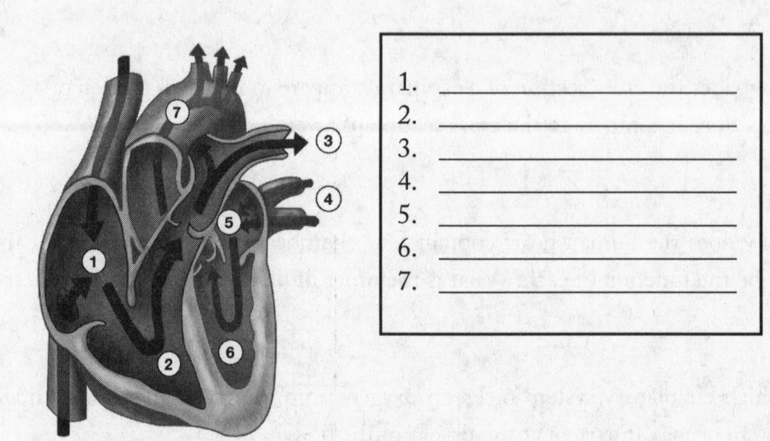

The Respiratory System and the Exchange of Gases

- Capturing oxygen and removing CO_2 occurs in the alveoli of the lungs. The lungs are composed of highly branched passages—the bronchi and bronchioles—to maximize the surface area available for gas exchange with the atmosphere. Structures that move gases from inside the body to outside include the nose, pharynx, and trachea.
- Air moves into and out of the lungs due to changes in air pressure during the breathing (respiratory) cycle. As the diaphragm expands (moves down), the chest volume increases so the internal air pressure decreases, drawing air into the body. During exhalation, the diaphragm relaxes, decreasing chest volume and increasing air pressure, and emptying the lungs.

- Respiration involves gas exchange—exchanging gases in the blood with the outside world through the lungs and exchanging gases between the blood and the interstitial fluid of the tissues. Both processes are driven by diffusion down concentration gradients.
- Oxygen and carbon dioxide are transported to and from tissues by the red blood cells. Oxygen travels bound to hemoglobin, whereas most of the carbon dioxide is transported as bicarbonate (HCO_3^-) within the red blood cell.

THINK ABOUT IT

1. How do humans breathe, and what muscles are involved?

2. In which structure within the lungs do red blood cells pick up oxygen? What molecule carries oxygen in the blood?

3. Why does blood need to carry carbon dioxide (CO_2)? How does it do this?

4. When spread out, the surface of alveoli in the human lung is said to approximate that of a tennis court. Why do you think this surface area needs to be so large?

5. Place the following terms in the correct columns of the table: high CO_2 concentration, low CO_2 concentration, high O_2 concentration, low O_2 concentration.

Lung	Pulmonary Capillaries

6. In the following figure, label these structures: larynx, bronchioles, esophagus, diaphragm, trachea, and lung.

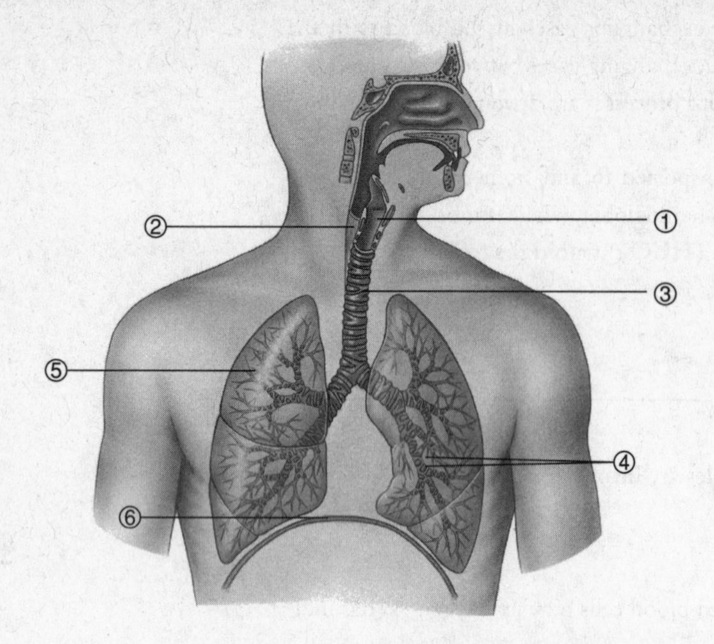

1. ____________
2. ____________
3. ____________
4. ____________
5. ____________
6. ____________

The Digestive System

- Components of the digestive system make small molecules out of big molecules, so that amino acids, fats, simple sugars, vitamins, and minerals diffuse into the circulation and are delivered to the tissues for use in synthetic reactions.
- Digestion consists of six steps. The first two, ingestion and mechanical degradation, occur within the oral cavity, pharynx, and esophagus. Food enters the mouth and is crushed and torn by the teeth. Saliva lubricates, and thus facilitates, the mechanical breakdown of the food. Enzymes within saliva begin the digestive process. The slurry of food now passes through the pharynx to the esophagus to make the trip to the stomach. The esophagus pushes the food down by waves of strong muscle contractions.
- The third and fourth steps of digestion occur within the stomach, where the macerated food is mixed with acid and enzymes, forming chyme. Gastric glands secrete a mixture of hydrochloric acid and the enzyme pepsin to facilitate breakdown of the proteins in food particles into amino acids. Muscular contractions of the stomach mix the chyme and gastric juice, accelerating the degradation.
- As the chyme enters the small intestine, additional enzymes secreted by the pancreas and bile secreted by the liver and gall bladder enter the digestive tract to complete the process of breakdown. The fifth step of digestion, absorption of small molecules into the capillaries and lymphatics, occurs almost exclusively in the small intestine.
- The final step of digestion, excretion of waste products, occurs through the action of the large intestine. Water is absorbed by the large intestine, compacting the waste products, which are eliminated by the rectum, a muscular chamber at the end of the large intestine that expels the fecal material.
- Different types of foods are degraded in different locations within the digestive tract. Carbohydrate digestion begins in the mouth and is completed in the small intestine; minimal degradation takes place in the stomach. Protein degradation is primarily accomplished by the gastric juice of the stomach. Fats are not digested until they enter the small intestine, where they are solubilized by bile salts and now serve as suitable substrates for the pancreatic lipase.

THINK ABOUT IT

1. Put the following steps in the correct order: digestion, ingestion, excretion, absorption.

2. In the context of the digestive system, briefly explain what each of the terms in question 1 means.

3. What kind of digestion occurs in the oral cavity?

4. What kind of teeth would you expect a cow to have? What about a wolf? (Think about Little Red Riding Hood.)

5. What compartments are connected by the esophagus?

6. What is the function of pepsin in the stomach? What about the function of hydrochloric acid?

7. Put the following sections of the small intestine in the correct order: jejunum, ileum, duodenum.

   stomach____ large intestine

8. What two important processes occur in the small intestine?

9. What does the gallbladder secrete? Where is this substance made, and what is its function?

10. What is the relationship between the colon and the bacteria it contains?

11. What happens to most of the water entering the digestive system each day?

12. Why should you be more careful about the amounts of vitamins A and D that you ingest than vitamin C?

13. Describe the digestion of a cheeseburger with lettuce on a bun. (You might want to divide the parts of a cheeseburger into their biochemical groups—fats, proteins, etc.—and then consider their fate that way.)

The Urinary System

- The urinary system—two kidneys, a bladder, and connecting structures—regulates fluid homeostasis. In addition to eliminating waste products, the kidneys maintain a constant blood volume and electrolyte concentration in the body.
- Kidneys remove waste products from the blood in the nephrons. Blood enters the Bowman's capsule under pressure, forcing water and small molecules, including the waste product urea, out of the blood and into the proximal tubule. As this filtrate travels through the tubule, $Na+$, amino acids, and sugars are actively pumped out of the tubule back into the blood, and the $Cl-$ ions and water follow to maintain osmotic and electrical balance. Water continues to be removed from the filtrate in the distal part of the tubule and the loop of Henle, as $H+$ and $K+$ ions and toxins move in. By the time the filtrate arrives at the collecting duct, it has become a concentrated solution of urea called urine.
- Urine travels to the urinary bladder through the ureters. The bladder holds up to 800 milliliters (mL), although the brain begins to receive signals from the stretch receptors signaling a need to urinate at 200 mL. Urine passes out of the bladder through the urethra, which has bands of smooth muscle (sphincters) at the top and bottom to control the flow of urine out of the bladder. Only the lower sphincter is under voluntary control.

THINK ABOUT IT

1. Organize the following terms with respect to the production and elimination of urine: urethra, kidney, ureter, bladder.

2. Circle the letter of each item that describes a function of the urinary system:

   a. regulate blood volume
   b. maintain ion balance
   c. control pH balance
   d. discharge waste

3. What substances enter and leave the kidney, and through which structures?

4. Why is it necessary that the kidney be capable of reabsorption?

5. When is urine actually produced?

6. Label this diagram of a typical nephron. What occurs in each portion of the nephron?

1 ______
2 ______
3 ______
4 ______
5 ______
6 ______
7 ______

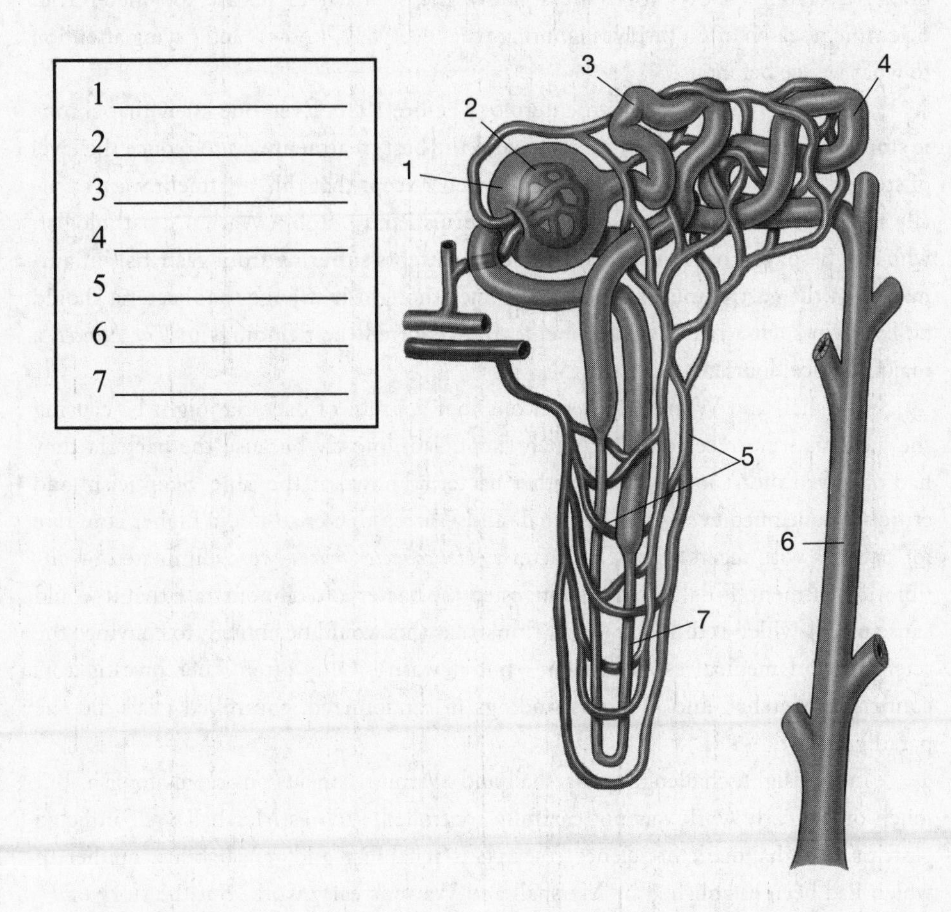

7. What are kidney stones?

8. Why do you think more females than males experience urinary bladder infections? (Think about the distance a bacterium must travel from the outside to the bladder in males and females.)

9. Compare and contrast the functions of the external and internal sphincter.

10. How can failure to urinate on demand cause incontinence?

Cool Fact: Learning to Expect the Unexpected

Ulcers caused by bacteria? Ridiculous. Everyone "knows" that organisms cannot survive at pH 1.4 (the pH level inside the stomach) or in the presence of pepsin. Everyone "knows" that ulcers are the result of too much acid destroying the mucosal membrane. Everyone "knows" that stress causes the stomach to release too much acid. Scientific research often involves ignoring what everyone "knows" and paying attention to what we see before us.

Such was the state of gastroenterology before 1980. Everyone knew that chronic stomach ulcers were caused by stress, and the best treatment was to reduce the level of stomach acid by changing the patient's diet. Except that this treatment wasn't usually successful. Enter Barry Marshall, an internist, and J. Robin Warren, a pathologist, who saw S-shaped bacteria in biopsies from patients suffering from gastritis (inflammation of the gastric mucosa). Although they thought it strange that bacteria should be living in such a harsh environment, they reported their findings in *The Lancet*, a major medical journal.

Marshall and Warren's observations that a strain of bacteria might be causing the gastritis were received with much skepticism, mostly because the bacteria they had observed didn't look like any other bacteria known at the time. Skepticism and criticism continued even after Marshall and Warren demonstrated a higher cure rate for patients with ulcers when the bacteria, *Helicobacter pylori*, were eliminated by antibiotic treatment. Finally Marshall ingested the bacteria to demonstrate that it would cause ulcers, which it did. You might think that this would be enough to convince the scientific and medical establishment—but it wasn't. Only after other investigators confirmed Marshall and Warren's findings in randomized, controlled trials did the paradigm shift.

In hindsight, challenging Marshall and Warren's conclusions seems foolish. But much of the early work was not carefully controlled, such as Marshall's self-inflicted gastritis. The hallmark of science research is reliability and repeatability, neither of which had been established by Marshall and Warren's early work. But the story of *H. pylori* and ulcers can stand as a reminder that identifying the source of an observation often forces us to ignore what we "know" to be true.

Chapter 28

TESTING, TESTING

Create a web:

1. blood, plasma, leukocytes, platelets, erythrocytes, hemoglobin, phagocytosis, immune system, oxygen transport
2. stages of digestion: mechanical processing, secretion, ingestion, digestion, excretion, absorption
3. food, elimination, circulation, excretion, metabolic waste, water, digestion, absorption

Make a flowchart: Trace the path of an oxygen molecule through the body.

Vocabulary Review

Place the following terms in the appropriate columns of the table. Some terms may be used more than once.

alveoli	erythrocytes	pancreas	urethra
blood volume	gas exchange	pH balance	venules
duodenum	intercostal muscles	platelets	
energy absorption	kidneys	regulation	

Cardiovascular	Pulmonary	Digestive	Urinary

PRACTICE TEST

Once you have finished studying this chapter, close your books, grab a pencil, and spend the next 15 to 20 minutes completing this practice test.

Compare and Contrast

For each of the following paired terms, write a sentence of comparison ("Both . . . ") and a sentence of contrast ("However, . . . ").

red blood cells/white blood cells

artery/vein

mechanical digestion/chemical digestion

alveoli/arterioles

ureter/urethra

closed circulation/open circulation

Matching

Match the following terms with their description. Each choice may be used once, more than once, or not at all.

a. water and nutrients move from the kidney tubule to blood vessels
b. release of urine
c. release of water, acids, etc., from the lining of the digestive tract
d. removal of water from the kidney filtrate
e. movement of small organic molecules from the intestine to the capillaries

___ Secretion
___ Absorption
___ Reabsorption
___ Concentration
___ Excretion

Short Answer

1. Explain the role of capillary beds in maintaining homeostasis.

2. How does oxygen move from air to blood?

3. What role does the liver play in digestion?

4. One of the major functions of the kidney is secretion. Explain the importance of secretion.

Multiple Choice

1. How does blood move through the veins from the body back to the heart?

   a. by the force of blood pressure
   b. heart contractions drive it

c. the squeezing of muscles, moving the blood through the veins
d. valves in the veins prevent backflow
e. c and d are both important in movement of blood through veins

2. Which of the following is not a part of the large intestine?

a. ileum
b. cecum
c. colon
d. rectum
e. a and b

3. Where does filtration occur within the nephron?

a. loop of Henle
b. Bowman's capsule
c. collecting duct
d. distal tubule
e. proximal tubule

4. Which chamber of the heart receives blood from the lungs?

a. right atrium
b. right ventricle
c. left ventricle
d. left atrium
e. a and d

5. Which solid element of blood is involved in clotting?

a. red blood cells
b. white blood cells
c. platelets
d. erythrocytes
e. leukocytes

6. Which compartment of the digestive tract has the lowest pH?

a. oral cavity (mouth)
b. esophagus
c. stomach
d. small intestine
e. large intestine

7. White blood cells

a. assist in blood clotting
b. transport oxygen

c. remove carbon dioxide from the blood
d. are active in defending the body against invaders
e. are red blood cells that lack melanin

8. The left ventricle is larger and stronger than the right ventricle because

a. the left is the first chamber that the blood enters as it returns to the body.
b. the left is the chamber that must pump blood to the lungs.
c. the left is the chamber that must pump blood to the body.
d. all the structures on the left side of the body are larger than those on the right.
e. Both b and d are correct.

9. Which of the following is the site of exchange for oxygen and carbon dioxide?

a. pharynx
b. trachea
c. bronchioles
d. alveoli
e. ravioli

10. Carbohydrate digestion begins

a. in the mouth.
b. in the stomach.
c. in the small intestine.
d. in the liver.
e. in the large intestine.

11. Which of the following is the first stop for absorbed nutrients after they leave the intestine?

a. pancreas
b. kidney
c. gallbladder
d. liver
e. prostate

12. All of the following are functions of the kidneys except

a. synthesizing vitamins.
b. controlling body pH.
c. water concentration in the body.
d. nutrient retention.
e. waste excretion.

13. A heart attack is caused by

a. a backup of bile in the liver.
b. a spasm during peristalsis in the duodenum.

c. lack of blood flow to cardiac muscle cells.
d. an autoimmune disease.
e. None of the above.

14. Ninety-two percent of plasma is made up of

a. red blood cells.
b. white blood cells.
c. platelets.
d. fibrinogen.
e. water.

15. Serafina is on a new health regimen. Especially concerned about her eyesight, she has decided to take megadoses of vitamins A and E every day. You explain to her that this is not a good idea, because

a. Vitamin K is much more important to eyesight.
b. Vitamins A and E are fat soluble and can accumulate to toxic levels.
c. Vitamins are broken down completely in the digestive system, so they have no effect when taken as supplements.
d. All the vitamins we need are produced within our bodies.
e. Both c and d are correct.

BUT, WHAT'S IT ALL ABOUT?

Here's a question to help you pull together what you've learned so far using this text. If you don't remember how to attack this type of question, check out the example in Chapter 1.

People with "poor circulation" suffer a variety of problems. Their fingers and toes may get cold easily. They are easily tired out by climbing stairs. They may suffer more frequent infections. They may seem lethargic. How are these ills connected to circulation?

What do I do now?
Remember the drill—decide what the question is asking you to do, collect your evidence from this chapter (and the others you've studied) and write!

29

CHAPTER 29

An Amazingly Detailed Script: Animal Development

Basic Concepts

- Development proceeds from general to specific. Undifferentiated cells of the zygote differentiate into three layers of germ cells that continue to differentiate into specialized tissues and, eventually, organ systems.
- Developmental processes are controlled through a hierarchy of gene product interactions; gene products diffuse through the cells, affecting the course of development.
- The body is shaped as cells migrate to new locations, adhere to each other, and die out when necessary.

General Processes in Development

- Development begins after fertilization, when the zygote begins to divide, forming a ball of cells of roughly the same size—an embryo.
- In the first stage of development this zygote continues to divide without cell growth, so the cells become very small as a morula forms. The morula, which is a solid ball of cells, rearranges to produce a liquid-filled cavity surrounded by cells. This is the blastula, the defining feature of all animals. The blastula also develops a polarity; one end is referred to as the animal pole, the other, the vegetal pole.
- Gastrulation follows, in which the cells rearrange again to make three layers; an inner endoderm, a middle mesoderm, and an outer ectoderm. These cell layers contain undifferentiated cells (germ cells) that will eventually produce specialized tissues; endoderm becomes the internal organs, mesoderm the bone and muscle, and ectoderm the nervous system and skin.
- The third phase of development, organogenesis, begins with the development of the nervous system. The notochord is a support structure that sends signals to the surrounding ectodermal tissue to develop into a neural tube.
- At the same time, mesodermal cells become somites on either side of the neural crest. These cells will become bones and muscle.

THINK ABOUT IT

1. Name and describe the three phases of development by filling in the following table:

Name of Stage	Description	Function
Cleavage		
	Cell movements and migrations	
		Formation of specialized organs

2. For each of the following structures, use the letters C (ectoderm), M (mesoderm), or N (endoderm) to explain their probable origin.

   ___ bunny fur ___ fish guts ___ body-builder muscles

   ___ your teeth ___ calf liver ___ your brother's skull

3. a. What phase of development is analogous to taking bread dough and dividing it again and again (and again and again)?

   b. During this process, does the total mass change?

   c. Does the mass of each piece change?

Factors Underlying Development

- Development is controlled by proteins called morphogens that activate genes to produce their products, so morphogens are transcription factors. The concentration gradient of a morphogen determines position information.
- Development is directed in a general-to-specific pattern by a hierarchical pattern of gene activation. The first genes activated specify a general position pattern. Subsequent genes fine-tune the position, and still later, additional genes direct the development of specific structures.
- Embryonic cells are totipotent; until the eight-stage cell of development, these cells can be directed to become any cell in the body. As development progresses, the fate of groups of cells becomes predictable. Although these cells are said to be "determined," their fate may change under the influence of morphogens. Cells that are no longer susceptible to the action of morphogens are "committed"—their fate has been determined and is not changeable.

THINK ABOUT IT

1. Use dots to represent the distribution of bicoid mRNA in a polar *drosophila* embryo.

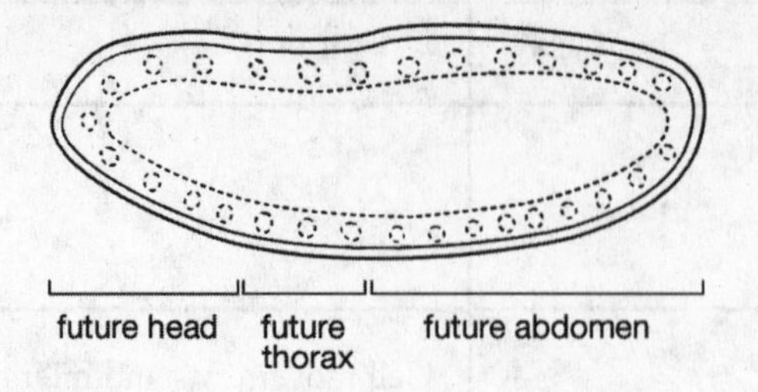

2. What are morphogens, and how do they work?

3. What process in development is analogous to a director assigning roles in a play? What role does this process play in development?

4. a. Place the following terms in the correct chronological order in the life of a cell: determination, totipotential, commitment.

   ______________ → ______________ → ______________

   b. What does it mean to say that a cell is determined?

5. The cloning of Dolly the sheep was groundbreaking because she developed from what kind of cell?

Shaping The Body

- Beyond the action of genes, the behavior of the cells of the developing embryo determines the shape of the body. Cells can move, adhere to each other, and die at the appropriate signal; these capabilities are essential to determine body shape.
- Cell death works to create spaces to create structure. Cells are not removed by other cells; rather, they self-destruct according to a predetermined program.
- Development does not end at birth—significant changes also take place in the body at puberty and menopause.

THINK ABOUT IT

1. Why does gastrulation require cell movement? Why would changes in cell-to-cell adhesion be involved?

2. What is the developmental process involving:

   a. migration of cells from one location to another?

   b. cells that anchor themselves before movement?

3. Development does not end at birth, but continues throughout the lifetime of the organism. Why?

4. Give a specific example of post-embryonic development.

Cool Fact: Hedgehogs and Limb Development

How can diffusing morphogens tell one set of amorphous, undifferentiated tissues to become a right hand and another a left? What signals tell embryonic cells to put digits on the far end of a limb and not sprout out of the shoulder or hip? Apparently, these decisions are directed by the product of a single gene, called *Sonic hedgehog.* (Yes, it's named for a video game character. Biologists *do* have a sense of humor.)

Ectoderm tissue, which will eventually become skin, guides the positioning of the muscles and tendons. Certain cells at the tip of an embryonic limb bud, the apical ectodermal ridge, control the order of structures within a limb—fingers at the far end, humerus bone near the shoulder. Whether these fingers are suitable for a right hand or a left hand is determined by another set of ectodermal cells, called the zone of polarizing activity (ZPA). In both cases, the gene encoding the morphogen is *Sonic hedgehog.*

Sonic hedgehog protein is expressed in the ZPA and is responsible for establishing the anterior-posterior and distal-proximal development of limbs. When ZPA cells from one side of an embryo's limb bud are transplanted to the opposite side, an inverted limb forms. Extra digits can be formed if the ZPA cells from one limb bud are transplanted onto another. If scientists delete the gene *Sonic hedgehog*, then the resulting animals have shortened and disorganized limbs. But these animals also have only one eye and abnormal brain development—clearly, *Sonic hedgehog* has a larger role than just directing where fingers are going to grow.

Identifying *Sonic hedgehog* as the key player in development of multiple tissues provides a target for scientists interested in understanding a variety of birth defects. Ultimately, our increased understanding of which genes control various elements of normal development will increase our understanding of diseases in which development goes wrong, such as cancer. Although it may seem strange that one gene could have so much "power," nature often uses one "tool" repeatedly and adapts its function to the situation. Understanding these adaptations is the next step in the process of defining the genetic control of embryonic development.

Chapter 29

TESTING, TESTING

Create a web: endoderm, mesoderm, ectoderm, hair, blood, liver, skeletal system, lungs

Make a flowchart: cleavage, organogenesis, gastrulation. What characterizes each stage?

Vocabulary Review

Place the following terms in the appropriate columns of the table. Some terms may be used more than once.

animal pole	ectoderm	mesoderm	notochord	
blastocoel	endoderm	neural crest	primitive gut	
blastula	invagination	neural tube	somites	zygote

Cleavage	Gastrulation	Organogenesis

PRACTICE TEST

Once you have finished studying this chapter, close your books, grab a pencil, and spend the next 15 to 20 minutes completing this practice test.

Compare and Contrast

For each of the following paired terms, write a sentence of comparison ("Both . . . ") and a sentence of contrast ("However, . . . ").

mesoderm/ectoderm

notochord/neural tube

blastula/gastrula

determined cells/committed cells

Matching

Match the following terms with their description. Each choice may be used once, more than once, or not at all.

a. Morphogen
b. Induction

c. Fertilization
d. Organogenesis
e. Gastrulation

____ conversion from haploidy to diploidy
____ notochord and neural tube develop
____ three-layered embryo develops
____ one tissue brings about the development of another tissue
____ a diffusible substance whose concentration affects development in that area

Short Answer

1. Explain the process of induction.

2. What are stem cells?

3. What is the difference between a zygote and an embryo?

4. How do morphogens affect development?

Multiple Choice

1. Development moves from the ____________ to the ____________.

   a. ridiculous; sublime
   b. animal pole; mineral pole
   c. top; bottom
   d. right; left
   e. general; specific

2. A committed cell

   a. has several options regarding its developmental fate.
   b. cannot reverse its developmental fate.
   c. is very dedicated to its job.
   d. is found in the endoderm.
   e. None of the above.

3. A cell that has the potential to develop into any cell type is

   a. committed.
   b. differentiated.
   c. totipotent.
   d. omnipresent.
   e. magnificent.

4. A blastula

   a. results from cleavage.
   b. is a ball of cells with an internal cavity.
   c. a and b.
   d. is shaped like a mulberry.
   e. has three germ layers.

5. A gastrula

   a. is a hollow ball of cells.
   b. results from cell movements.
   c. is shaped like a mulberry.
   d. has three germ layers.
   e. b and d.

6. During development, cell A causes cells B and C to develop into specific struc tures. This process is known as:

   a. invagination.
   b. imbibition.
   c. inhibition.
   d. induction.
   e. incarceration.

7. Which of the following is NOT a developmental process?

   a. puberty
   b. cleavage
   c. aging
   d. cellular respiration
   e. organogenesis

8. The early embryonic structure that resembles a berry-like ball of cells is known as the

   a. blastula.
   b. gastrula.
   c. morula.
   d. archenteron.
   e. berryball.

9. Development is controlled by the interaction of

   a. sand and sea.
   b. animals and vegetables.
   c. proteins and lipids.
   d. genes and proteins.
   e. cells and minerals.

10. Cells in which of the following stages would be the easiest to use as a source for cloning?

    a. determined
    b. committed
    c. driven
    d. totipotent
    e. totally confused

11. Serafina needs your help. She is trying to prepare for an exam in her biology class and is having trouble with the stages of development. She shows you her study notes and asks you to correct them. Which of the following statements from Serafina's notes is a correct statement?

    a. The three layers of embryonic tissue are ectoderm, mesoderm, and wrinklederm.
    b. Organogenesis comes after gastrulation but before cleavage.
    c. Stem cells are uncommitted.
    d. The zygote stage occurs right before birth.
    e. The morula is a developmental stage in mulberries.

12. Your project in bio lab is to study the development of a bioengineered species, *Swimmerela fastus*. As you experiment with *fastus*, you realize that, similar to *drosophila*, there is a bicoid-like substance that controls local development through changes in its concentration. Molecules that affect development in this way are known as

    a. morphogens.
    b. antigens.
    c. pyrogens.
    d. antibodies.
    e. prions.

BUT, WHAT'S IT ALL ABOUT?

Here's a question to help you pull together what you've learned so far using this text. If you don't remember how to attack this type of question, check out the example in Chapter 1.

It has been said, "ontogeny (development) recapitulates phylogeny (evolution)." What does this mean?

What do I do now?

Remember the drill—decide what the question is asking you to do, collect your evidence from this chapter (and the others you've studied) and write!

30

CHAPTER 30

How the Baby Came to Be: Human Reproduction

Basic Concepts

- Reproduction begins when eggs and sperm fuse to create a zygote in an environment favoring development.
- A human male makes gametes (sperm) continuously through his lifetime; the sperm precursor cells, spermatogonia, generate not only sperm but also new spermatogonia.
- The female makes gametes (oocytes) once a month for 35 to 40 years. All of the eggs she will ever produce are present in a suspended state of development when she is born.
- Early human development proceeds rapidly through the stages of animal development, but fetal growth occurs more slowly.

Overview of Human Reproduction

- Monthly, one oocyte matures within a follicle in the female ovary under the influence of estrogen and progesterone. It is released to float down the uterine tube toward the ovary.
- In contrast, male sperm are produced continuously from spermatogonia under the influence of testosterone. Sperm mature within the epididymis and travel up the vas deferens and out the urethra upon ejaculation.
- Fertilization occurs when one of the millions of sperm swimming up the vagina and into the uterus penetrates the oocyte. The surface of the oocyte changes so that no other sperm may enter. The fertilized egg continues its transit toward the uterus, ultimately implanting itself into the uterine wall.

THINK ABOUT IT

1. Despite their many differences, male and female reproductive systems have two purposes in common. What are they?

2. What hormones influence the male and female reproductive systems, and where are they produced?

The Female Reproductive System

- Two processes occur within the female reproductive cycle—an oocyte matures and becomes available for fertilization, and a suitable environment is prepared for the zygote.
- If the egg is not fertilized, the uterine endometrium is sloughed off, causing menstruation.
- Follicle loss occurs continuously over a female's lifetime, so that by age 50 she may have fewer than 1,000 follicles remaining. This may be the trigger that causes the reproductive cycle to end (menopause).

THINK ABOUT IT

1. During ovulation, eggs are released from the ___________ into the ___________.

2. Where does fertilization usually occur?

3. What is the endometrium, and what is its role?

4. What is occurring during menstruation?

5. What is the function of the corpus luteum?

6. How many zygotes develop in a pregnancy with fraternal twins? How about in a pregnancy with identical twins?

7. What process leads to fraternal twins? Mike and Mary are a pair of fraternal twins. Which one is more likely to have twins in turn?

8. Use your text as a key when you label this figure with the following terms: uterus, zygote, accessory cells, oocyte, uterine tube, implantation, ovary.

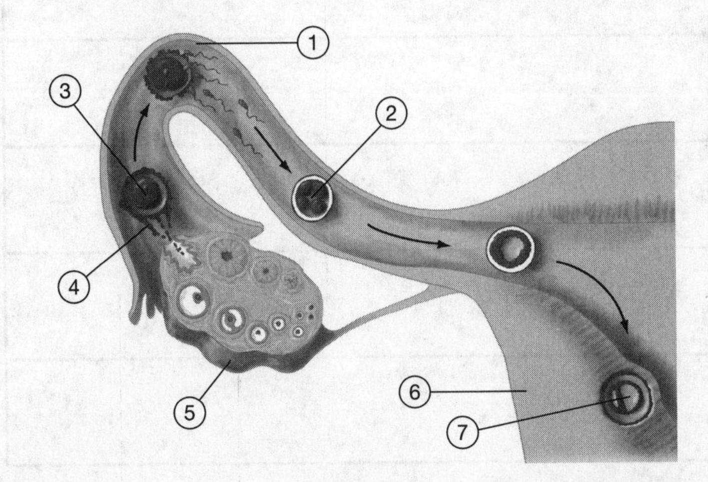

1. ___________
2. ___________
3. ___________
4. ___________
5. ___________
6. ___________
7. ___________

The Male Reproductive System

- Sperm mature within the seminiferous tubules of the testis. About 250 million new sperm are made each day. Mature sperm move into the epididymis for additional maturation and storage; about 2.5 months are needed for the complete maturation process.
- Sperm travel through the vas deferens to reach the urethra. Sperm are mixed with seminal fluid produced by the accessory glands, seminal vesicles, bulbourethral glands, and prostate gland before ejaculation. Sperm account for only 5 percent of the volume of the semen.
- A single spermatozoon enters the oocyte by digesting part of the outer layer surrounding the oocyte. An acrosome containing enzymes digests this layer after it becomes capacitated by the action of substances in the outer layer of the oocyte. Only one sperm succeeds in penetrating the oocyte; after penetration the membrane potential of the oocyte changes, blocking entry of additional sperm.

THINK ABOUT IT

1. Why don't males experience menopause?

2. A female normally releases one egg each month. How many sperm are released in each male ejaculation?

3. At each of the following structures, describe what is happening to the sperm.

Structure	Function
Testes	
Epididymis	
Seminal vesicle	
Prostate gland	
Urethra	
Bulbourethral gland	

4. What is the function of the sperm acrosome, and why is it necessary?

5. How is the electrical potential of the egg's membrane used as a defense against polyploidy? Why is this necessary?

Human Development and Birth

- Human fetal development is completed—meaning organogenesis is complete—in 12 weeks, the first trimester of the pregnancy. During the second and third trimester, the fetus grows, increasing in length during the second trimester and in weight during the third.
- The human blastocyst forms two structures during early embryonic development—the embryo itself and the placenta, the organ that creates a network of maternal and fetal blood vessels to supply the fetus with nutrients.
- Regular contractions of the uterus at about 38 weeks post-fertilization signals the beginning of the birth process (labor). Birthing proceeds in three phases—first, the cervix must open to allow the baby to pass out of the mother; second, the baby is pushed out by contractions of the uterine muscle; and third, the placenta is expelled.

THINK ABOUT IT

1. What is the embryonic period of human development, and how long does it last?

2. Complete the following table including the major components of the human blastocyst and what structures they give rise to:

Part of Blastocyst	Gives Rise to:

3. What is surfactant, and how does its production affect premature babies?

4. In the human embryo, what significant stages occur at each of the times listed in the following table? (Stages: organogenesis, gastrulation, blastula formation, implantation)

Time	Stage
5 days	
7–10 days	
16 days	
4th week	

5. Describe the three phases of the human birth process.

6. What is the afterbirth?

Cool Fact: Alien?

A baby in the womb is physically and genetically distinct from its mother; half of the baby's genes are paternal and are producing proteins like those present in the father. So why doesn't the mother's immune system recognize the fetus as a foreign invader and try to eliminate it?

Fetal tissue comes in contact with maternal blood within the placenta, the organ responsible for delivering nutrients and oxygen to the fetus. Paternally derived proteins can also be delivered into the maternal circulation in the placenta, and maternally derived immune cells and molecules can cross into fetal circulation. Although we'd expect the maternal immune system to be activated by the paternal antigens and send in immune cells to eliminate the invader, this happens only rarely. In a condition called recurrent spontaneous abortion (RSA), the maternal immune system does produce high levels of natural killer (NK) cells, which do attack the fetus and cause a spontaneous abortion (miscarriage). Apparently, NK cell activity must be significantly decreased in a normal pregnancy, but how this control happens is still unknown.

Two treatments that help prevent maternal rejection of the fetus in women experiencing RSA include immunization with paternal leukocytes and intravenous immunoglobulin therapy. Both treatments work by depressing NK cell activity, allowing the pregnancy to run its full course. Both treatments involve introducing new antibodies into the mother's circulation, but how or why these immunoglobulins protect the fetus are the subject of current research. Even though we don't know why the treatments work, we know they do work—and for parents anxious to have a child, that is all that matters.

TESTING, TESTING

Create a web: trimesters 1, 2, and 3; organogenesis; gastrulation; cleavage; kicking; "breathing"; birth

Make a flowchart: Include trimesters 1, 2, and 3; what are the major changes during each stage?

Vocabulary Review

Place the following terms in the appropriate columns of the table. Some terms may be used in more than one column.

blastocyst embryo ovary semen spermatocyte

conception oocyte placenta seminiferous tubule zygote

Female Reproductive System	Male Reproductive System	Human Development

PRACTICE TEST

Once you have finished studying this chapter, close your books, grab a pencil, and spend the next 15 to 20 minutes completing this practice test.

Compare and Contrast

For each of the following paired terms, write a sentence of comparison ("Both . . .") and a sentence of contrast ("However, . . .").

sperm/semen

uterine tube/urethra

epididymis/corpus luteum

oocyte/zygote

Matching

Match the following terms with their description. Each choice may be used once, more than once, or not at all.

a. Oogenesis
b. Atresia
c. Ovulation
d. Conception
e. Bulbourethral gland

_____ the natural degeneration of follicles
_____ the release of an oocyte from a follicle
_____ the formation of female gametes
_____ accessory structure for semen production
_____ fusion of two haploid nuclei

Short Answer

1. What is atresia?

2. Why are the testes located outside of the body?

3. What is the difference between identical and fraternal twins?

4. Why is lung function such an important factor in premature babies?

Multiple Choice

1. The corpus luteum develops from

   a. the endometrium.
   b. an immature follicle.
   c. a follicle after it has ruptured and released an oocyte.
   d. the uterus.
   e. None of the above.

2. What is cervical dilation?

   a. dilution of the amniotic fluid
   b. contraction of the muscles of the uterus during labor
   c. the opening of the cervix so that the baby can pass through
   d. constriction of the cervix, which holds the fetus in the uterus during development
   e. None of the above.

3. Conception occurs in the

   a. uterus.
   b. seminiferous tubules.
   c. ovary.

d. follicles.
e. uterine tubes.

4. What is polyspermy?

a. ejaculation of multiple sperm
b. fertilization of the oocyte by more than one sperm
c. fertilization of more than one oocyte at the same time
d. formation of four sperm from one precursor cell
e. fertilization in a laboratory dish

5. A human blastocyst is equivalent to

a. a gastrula.
b. an inner cell mass.
c. a trophoblast.
d. a blastula.
e. a morula.

6. What does a fetus start to breathe during the third trimester?

a. air
b. blood
c. amniotic fluid
d. water
e. lymph

7. How many eggs and sperm are required to produce fraternal twins?

a. one egg, one sperm
b. one egg, two sperm
c. two eggs, one sperm
d. two eggs, two sperm
e. two eggs, over easy

8. Ninety-five percent of semen is comprised of

a. the egg plus its surrounding cells.
b. sperm.
c. the fertilized egg and its support structures.
d. support materials for the sperm.
e. Navy personnel; the other 5 percent are Marines.

9. Implantation of the blastocyst occurs

a. just before conception.
b. one week after fertilization.
c. at one month gestation.

d. just before birth.
e. implantation is not a developmental process.

10. Why are contractions necessary during birth?

 a. so the mother will know that the baby is about to arrive
 b. so the vagina will expand to accept the baby
 c. so the ovaries will know when to resume oocyte production
 d. so the cervix will expand to let the baby through
 e. no pain, no gain

11. Serafina is pregnant! She and her husband, Lee, are thrilled to be starting a family. But Serafina is worried. She hasn't mentioned it to Lee, but she hasn't felt the baby move or heard its heartbeat. You tell her not to worry; this is natural because

 a. Pregnant women often have a decrease in sensitivity of their hearing.
 b. Babies rarely move before the third trimester.
 c. She is only in her first trimester, and the baby has not formed all of its structures yet.
 d. The heart does not need to start beating until birth.
 e. You always suspected that she would give birth to an alien anyway.

12. The most likely explanation for the observation that human males release approximately 200,000,000 sperm in each ejaculate is that

 a. it requires enzymes from that many sperm to breach the fluids of the vagina.
 b. it allows males to overcome competition from other males seeking to fertilize the female's egg.
 c. since each sperm carries only one chromosome, multiple sperm are required to fertilize each egg.
 d. the unused sperm will form the protective layer around the developing embryo.
 e. fertilization is a team sport.

BUT, WHAT'S IT ALL ABOUT?

Here's a question to help you pull together what you've learned so far using this text. If you don't remember how to attack this type of question, check out the example in Chapter 1.

Implantation of the embryo in the uterine wall is described as an "invasion" in the textbook; the embryo inserts itself into maternal tissue and links up with the maternal blood supply. The embryo, possessing genes from both parents, is different from the mother – it is "not-self" from the mother's perspective. Why isn't the embryo attacked by the mother's immune system?

What do I do now?
Remember the drill—decide what the question is asking you to do, collect your evidence from this chapter (and the others you've studied) and write!

CHAPTER 31

An Interactive Living World: Populations and Communities in Ecology

Basic Concepts

- Ecology studies the interactions of organisms with each other and with the physical environment.
- Growth of organism populations would be exponential if not limited by environmental resistance.
- Species within a community interact with each other in their pursuit of resources. These interactions may be competitive or noncompetitive, and helpful, harmful, or neutral.
- Communities devastated by natural forces, or human abandonment, will regenerate in a predictable pattern of succession.

Population Dynamics

- Populations can grow arithmetically or logarithmically. Arithmetic growth shows a constant rate of population increase over time, whereas logarithmic growth starts out slowly but increases very rapidly because the rate of growth is proportional to the number of individuals in the population.
- In the real world, factors in the environment such as predators, disease, and limited resources prevent logarithmic population growth.
- Most populations grow logistically; that is, the rapidly accelerating growth rate that occurs when the population is small slows, stabilizes, and eventually plateaus as a balance is achieved between the number of organisms in an area and the level of available resources.
- The intrinsic growth rate of a species depends, in part, on whether it is a long-lived, density-dependent equilibrium species (*K*-selected) or a short-lived, density-independent, opportunistic species (*r*-selected).
- Changes in population size, especially human populations, are also affected by the proportion of the population at or approaching the reproductive stage and by the movement of individuals into and out of the population.

THINK ABOUT IT

1. Label the following diagram with these terms: population, ecosystem, biosphere, community.

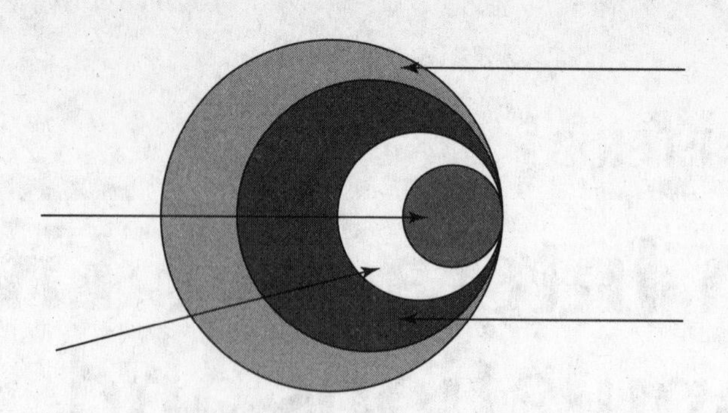

2. Classify each of the following as a population (P), community (C), or ecosystem (E).

    ___ a. the students in your biology class
    ___ b. the birds, bees, and flowers in your garden
    ___ c. the soil, fertilizer, and crops in a farmer's field
    ___ d. the pumpkinseed sunfish in a pond
    ___ e. the pumpkinseed sunfish, plankton, and plants in a pond

3. What are the three parameters involved in population dynamics?

4. Which of the following populations could be directly counted, and which would require an estimate? For those that require an estimate, how would you do it?

    a. the number of zooplankton in a lake
    b. the number of cacti in the Sonoran desert
    c. the number of penguins at the New England Aquarium
    d. the number of wild blueberry bushes on a small island in Maine
    e. the number of grizzly bears in Banff National Park

5. Label these two curves as exponential or arithmetic growth. Give an example of each type of population growth.

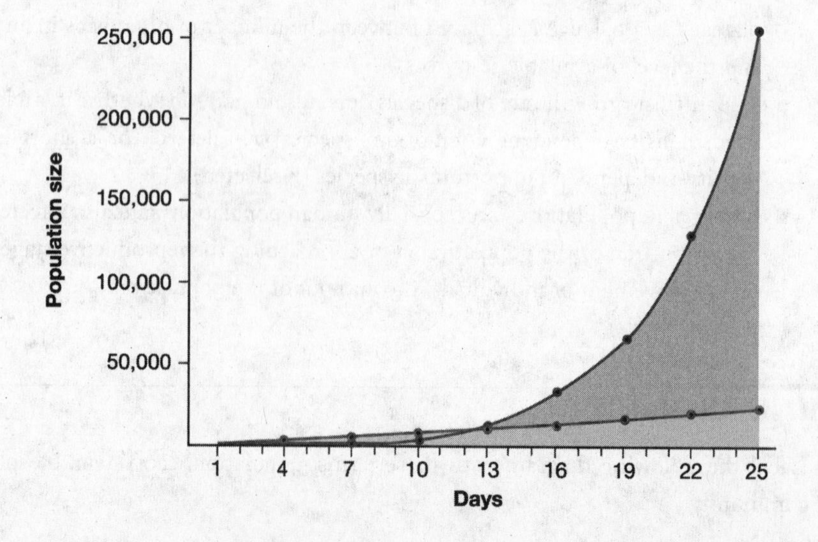

6. a. In the following spaces, plot an exponential and a logistic growth curve.

Exponential:

Logistic:

What accounts for the differences between the two curves?

b. A population of aphids inhabits a rose bush. Over time the population increases very rapidly until new aphids can no longer find enough places on the plant to sit and suck sap. At this point the population size levels off and remains steady until the first hard frost of autumn kills all the aphids. Which of the two curves in part (a) best describes the growth of this aphid population? Which term describes the hard frost and the lack of space on the plant?

7. Use the following data to calculate *r* for this population:

# individuals present 1/1/03	# births 1/1/03–1/1/05	# deaths 1/1/03–1/1/05
3000	350	285

8. Would you expect two populations of the same species in different locations to experience the same *K*?

9. Why do people study ecology at the level of a single population?

10. Which is more likely to occur in the real world, arithmetic or exponential growth? Why?

11. A population of chimpanzees had 100 individuals in 1997. Fifteen individuals were born and eight died that year. What is the intrinsic rate of growth?

12. The yearly birth rate among a population of flamingos in western Canada is 12 percent. If the population has 200 individuals today, how many chicks would you expect to hatch during this year?

13. What are the advantages for a population that is *r*-selected rather than *K*-selected?

14. Of the following factors, which would have an impact on the population density in a *K*-selected population?

    a. disease
    b. decrease in temperature
    c. increase in food supply

15. Mark the following survivorship curves as late-loss, constant loss, and early loss. Which curves are characteristic of *r*- and *K*-selected populations? Give an example of a population that is typical of *r*- and *K*-selected populations.

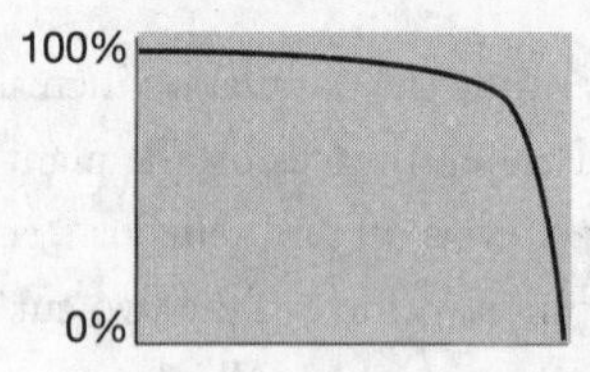

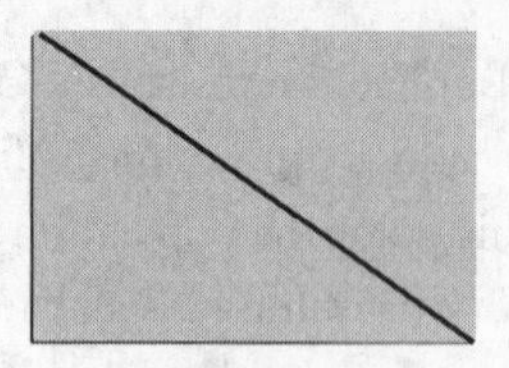
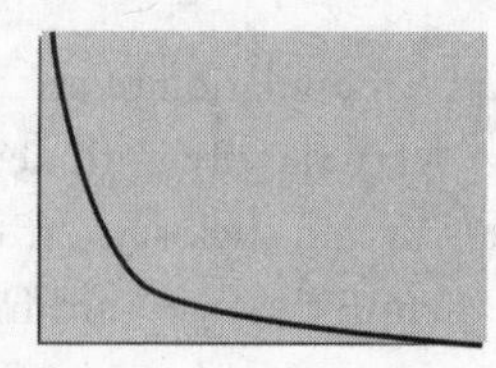

16. For each of the following populations, mark it as *r*- or *K*-selected:

    ___ dandelions that move into an untreated lawn and rapidly take over
    ___ beech trees that don't reproduce until age 20
    ___ elephants, which produce only one offspring at a time
    ___ cockroaches (with egg cases full of 100 eggs each)

17. a. Draw a population pyramid for a population past reproductive age and one for a population at or prior to reproductive age:

70–80 years ____
60–70 years
50–60 years
40–50 years
30–40 years
20–30 years
10–20 years
0–10 years ____

70–80 years ____
60–70 years
50–60 years
40–50 years
30–40 years
20–30 years
10–20 years
0–10 years ____

b. Which of these populations is likely to experience the most rapid population growth? Why?

18. In addition to birth and death rates, what factors influence population change?

19. Which do you think will have a greater impact on the survival of our species—the number of individuals or the amount of natural resources used by each individual? Explain your answer.

20. Name three density-dependent factors that could affect the carrying capacity of humans on Earth.

Species Interactions within the Ecological Community

- Loss of the keystone species significantly changes the population dynamics of the community.
- Keystone species are often the top predators within a community, but any species necessary for the dynamic equilibrium of species interactions may be a keystone species.
- Populations within a community may compete for the resources of an environment—food, shelter, and space. Since no two species can share the same resources without one species driving out the other, similar species coexist

within the community only when resources can be partitioned (used at different times).

- One species may serve as a resource for another, as in the interactions defined by parasitism and predation. Predator and prey animals keep each other's populations in equilibrium because their population cycles are linked.
- Predator-prey interactions exert evolutionary pressure such that the species involved may co-evolve. Prey species must develop avoidance or defense mechanisms, whereas predator species must evolve ways to overcome prey defense systems or use other species as food.
- Mutually beneficial interactions exist, some to the extent that the two species depend on each other for survival.

THINK ABOUT IT

1. What are the ecological dominants in a community?

2. What are different measures of biodiversity?

3. The number of individuals of many species has declined to dangerous levels, but many species are beginning to return to healthy numbers, due to conservation efforts and captive breeding programs. What aspect of biodiversity is in jeopardy because of this decline and expansion in numbers? What are the implications for the restored species?

4. Draw a community with five species. Which one is likely to be the keystone species?

5. What is the purpose of studying ecology at the community level?

6. You do two mixing experiments in an artificial pond with different species of plankton.

   Experiment 1: Mix species A and B.
   Experiment 2: Mix species C and D.

   At the end of experiment 1, you can find only plankton of species B in the pond, while at the end of experiment 2, you can find approximately equal numbers of individuals of species C and D.

   a. What is happening in each experiment?
   b. Which two species are likely to have the most similar niches?

7. Infection of humans by certain organisms results in a "chronic" long-term infection (e.g., malaria). Is this an example of parasitism or predation?

8. If *all* poisonous plants on a fictional island had red flowers, what kind of mimicry would be at work?

9. Your pet dog was once stung by a bee and now avoids all black-and-yellow flying insects, even ones that don't sting. What form of mimicry is protecting the harmless insects from your dog's attentions?

10.

Both species hurt | Both species helped

Place the following terms on the bar above, in the appropriate order: Batesian mimicry, Mullerian mimicry, predation, commensalism, parasitism.

11. What process allows two species to coexist even as they undergo competitive exclusion?

Succession In Communities

- The process of succession begins with a condition where little to no biotic components of an environment remain after a disaster, or in the case when a community disturbed by humans has been abandoned.
- Succession begins as pioneer plant species, carried by the wind or birds, begin to repopulate the area. Other plants succeed the pioneer species as the environment is changed by the earlier-arriving species.
- Succession produces a trend toward longer-lived species, increased biomass, and increased species diversity despite driving out earlier colonizing species.

THINK ABOUT IT

1. If plant growth was observed on the moon after an astronaut's visit, what type of succession would that represent?

2. Label each of the following as typifying primary succession (P), secondary succession (S), or climax community (C):

___ weeds growing through the cracked asphalt of an abandoned gas station

___ a mature, stable forest community

___ a farm field allowed to return to the wild
___ a plant community growing on the cooled lava of a Hawaiian volcano

3. What are climax communities?

Cool Fact: Predator and Prey in an Urban Environment

The white-tailed deer *(Odocoileum virginianus)* were once scarce inhabitants of deciduous forests in the northeastern United States; the population was almost driven to extinction at the beginning of the twentieth century. Hard to believe that the white-tail is now the dominant species of the forest.

Several factors enabled the deer populations to recover after being hunted to near extinction. Logging and clearing the land for crops eliminated many of the deer's natural predators (cougars, black bears, and wolves). Young trees populated spaces cleared by logging, providing abundant food for the deer. State wildlife agencies enforced laws protecting deer populations, allowing the herds to recover enough to support limited hunting. But the effect of 100 years of "protection" has produced large deer populations with no natural enemies—living next to suburban and urban human communities.

In human terms, these deer are costly. Deer-vehicle collisions cost motorists thousands of dollars annually. Deer also consume ornamental plantings, forcing homeowners to either erect very high (over 6 feet) fences or shoulder the costs of replacing the plants. If your sympathies still lie with Bambi and his family, think also of the "cost" of unrestricted deer browsing within the remaining forest lands; regenerating populations of hardwood trees can become stunted by overbrowsing, and the understory plants can completely disappear, eliminating most of the biodiversity in a forest.

So what's to be done? Reintroducing the predator species is not a practical solution in the human-population-dense Northeast Corridor. Many communities have organized controlled hunting, usually hiring sharpshooters, to reduce the deer population to a level that can be supported by the ecosystem. But such strategies often create controversy when community residents realize that people will be hunting around their homes and schools. Contraception methods have been developed for deer populations; these involve injecting the deer with a protein that causes the deer's immune system to develop antibodies against a developing embryo. Despite being more humane than killing the animals, immunocontraception does not decrease the population size, so overbrowsing is still a problem. Besides, the contraceptive is expensive, requires repeated applications, and is most often administered by shooting the animal with a dart gun—so people are still firing guns in residential areas.

There are no simple answers to the question of how to support deer and people in the same space. What our experience should show us, though, is how elegantly balanced are predator-prey relationships and the consequences of upsetting the balance by removing the predator.

TESTING, TESTING

Create a web: *r*-selected, *K*-selected, carrying capacity, environmental resistance, logistic growth, exponential growth, equilibrium species, opportunistic species

Make a flowchart: Include the scales of life (physiology, populations, communities, ecosystems, biosphere). What is happening to you at each of these levels as you read this question?

Vocabulary Review

Place the following terms in the appropriate columns of the table. Some terms may be used in more than one column.

biodiversity	groups of individuals	*K*	
biosphere	groups of species	keystone species	
carrying capacity	intrinsic rate of growth	*r*	study of the home
ecosystem	J-curve	S-curve	

Community	Population	Ecology

PRACTICE TEST

Once you have completed this chapter, close your books, grab a pencil, and spend the next 15 to 20 minutes completing this practice test.

Compare and Contrast

For each of the following paired terms, write a sentence of comparison ("Both . . .") and a sentence of contrast ("However, . . .").

competition/predation

primary/secondary succession

community/population

mutualism/commensalism

Matching

Match the following terms with their description. Each choice may be used once, more than once, or not at all.

a. the elimination of one species by the more effective use of a common resource by another species
b. false message sent to one species by another
c. the study of the natural "home"
d. growth of a population limited by some factor
e. mutual evolution of two species, where the adaptations of one influence the adaptations of the other

___ competitive exclusion
___ logistic growth
___ ecology
___ mimicry
___ coevolution

Short Answer

1. What is a survivorship curve?

2. What term would you use to describe a community interaction in which one species benefits while the other experiences no change in circumstances?

3. What two processes play a role in most cases of primary succession?

4. How does predation differ from parasitism?

Multiple Choice

Circle the letter that best answers the question.

1. Carrying capacity is.

   a. the rate of population increase.
   b. the difference between the birth rate and the death rate.
   c. always the same for a given species.

d. the maximum population density of a given species in a given geographical location.
e. exponential growth of a population.

2. Immigration will tend to:

a. stabilize population size.
b. decrease population size.
c. increase population size.
d. cause a *K*-selected species to become *r*-selected.
e. None of the above.

3. Which of the following is true about keystone species?

a. They are always the largest species in a community.
b. They are always the top predator.
c. They never interact with other species.
d. They maintain the balance of power within the community.
e. All of the above.

4. What is a climax community?

a. the community at the top of a mountain
b. the community that first appears after a natural disaster
c. the stable community achieved at the end of a succession process
d. the community at its carrying capacity
e. the community without its keystone species

You are doing research on the island, Ecophilia. The island is very large, with some habitats and ecosystems heretofore unseen. In one particular habitat, your studies indicate that environmental conditions are relatively unstable. For example, inhabitants experience sudden torrential rainstorms followed by hot, sunny days.

5. The first species you encounter, *Mickymousiania*, is a small, mammal-like creature that seems to produce many offspring per gestation period. When their offspring are born, the parents seem to ignore them in favor of preparing for the next litter. From this you conclude that

a. This species is likely to be *r*-selected.
b. This species is likely to be *K*-selected.
c. This is an equilibrium species.
d. This is an opportunistic species.
e. Both a and d are correct.

6. In another habitat on Ecophilia, you discover *Dumbosiania*, a larger, less-numerous relative of Mickymousiania. Due to its slower growth and decreased reproductive rate, you conclude that the population growth of Dumbosiania is probably

a. opportunistic.
b. *r*-selected.
c. density-dependent.

d. density-independent.
e. goofy.

7. Communities undergoing primary succession are

   a. stable and unchanging.
   b. not allowed to vote in some elections.
   c. found only near the equator.
   d. being engulfed by a nearby metropolis.
   e. started from bare rock.

8. When startled, a moth unfurls its wings to reveal spots that resemble owl eyes. This is an example of

   a. commensalism.
   b. parasitism.
   c. Batesian mimicry.
   d. Mullerian mimicry.
   e. predation.

9. Populations experiencing rapid, unchecked growth would have a growth curve that resembles the letter

   a. A.
   b. R.
   c. K.
   d. J.
   e. S.

10. Which of the following could be defined as the job description of a species?

    a. population
    b. habitat
    c. niche
    d. ecosystem
    e. want ads

11. A researcher wanting to learn more about the interactions between bees and the flowers they pollinate would study a

    a. species.
    b. population.
    c. community.
    d. ecosystem.
    e. biosphere.

12. A population that has stopped increasing in size due to limited resources has

    a. become extinct.
    b. maintained its fitness.

c. reached its carrying capacity.
d. had to move to new niche.
e. little to teach us about ecology.

BUT, WHAT'S IT ALL ABOUT?

Here's a question to help you pull together what you've learned so far using this text. If you don't remember how to attack this type of question, check out the example in Chapter 1.

Viruses, although not living things, do establish a relationship with their hosts similar to parasitism in that the viruses "feed" on their hosts by extracting the energy, materials, and machinery necessary to replicate and infect other hosts. Is it correct to represent a viral infection as a form of parasitism?

What do I do now?

Remember the drill—decide what the question is asking you to do, collect your evidence from this chapter (and the others you've studied) and write!

32

CHAPTER 32

An Interactive Living World: Ecosystems and the Biosphere

Basic Concepts

- Ecological systems are functioning units composed of living organisms and the physical (abiotic) environment.
- Energy, needed to fix carbon in biogeochemical cycling, flows through the ecosystems and is neither created nor destroyed along the way—only transformed.
- Six biomes—ecosystems of varied productivity defined by their vegetation—are determined by the climate, the average temperature, and the amount of precipitation.

Ecological Systems

- Nonliving components—resources such as minerals, nutrients, and energy sources; and conditions such as climate—are key components of every ecosystem.
- Carbon and nitrogen are key minerals that cycle between the biotic and abiotic components of the ecosystem. Carbon cycles between biotic components, mostly plants, and abiotic components, such as coal, in the form of carbon dioxide gas (CO_2) Nitrogen is also fixed by living organisms in the form of proteins and nucleic acids, but it cannot be extracted from the abiotic world—where it exists as nitrogen gas (N_2)—except by bacteria.
- Water, essential to life, cycles between large bodies of water (such as lakes, the oceans, and the polar ice caps) and water vapor in the atmosphere.

THINK ABOUT IT

1. Mark the following components of an ecosystem as biotic (B) or abiotic (A):

   ___ water in a river
   ___ calcium-rich limestone bedrock
   ___ worms tunneling in the soil
   ___ fungi decomposing leaf litter

2. How do ecosystems differ from communities?

3. What are the elements required in the greatest amounts by living organisms?

4. What is the source of the increased CO_2 in the present-day atmosphere?

5. Complete the following sentence by circling the correct words and filling in the blanks:

   Plants generate/consume CO_2 by the process of ____________________, and animals generate/consume CO_2 through the process of ______________ ______________.

6. How is "stored" carbon in the wood of a tree released once the tree dies? (And can it be heard in a forest?)

7. Which macromolecules have nitrogen in their structure?

8. a. What are the starting and end products of nitrogen fixation?

   Starting material—

   End product—

   b. Why is nitrogen fixation a critical process?

9. a. What are the two processes by which water enters the atmosphere?

   b. What role does the Sun play in these processes?

10. Name two reservoirs (sources) of nitrogen for plants.

Energy Flow Through Ecosystems

- Living organisms collect and store energy temporarily; they interrupt the flow of energy from the Sun to its ultimate fate—dissipation as heat.
- Producer organisms, such as plants, capture energy through photosynthesis and make it available in the form of food for consumer organisms. Producer

and consumer organisms are connected by food webs known as trophic levels.

- Energy flowing from lower trophic levels (plants) to higher levels (secondary and tertiary consumers) declines by 90 percent for each step in the process. These large losses mean that organisms on the top tropic level (tertiary consumers) are rare.
- Climate affects energy flow, and climate is determined by the Earth's physical environment. Atmospheric circulation acts to distribute water around the globe, creating "bands" of relatively "wet" and relatively "dry" environments around the Earth. Changes in the Earth's temperature due to global warming could affect atmospheric circulation and precipitation patterns and thus change physical environments.

THINK ABOUT IT

1. Classify the following energy sources as ordered or disordered:

   ___ sugar
   ___ gasoline
   ___ heat
   ___ bread

2. a. Place the following organisms into the appropriate trophic level boxes below: blue jay, coyote, sunflower, house cat.

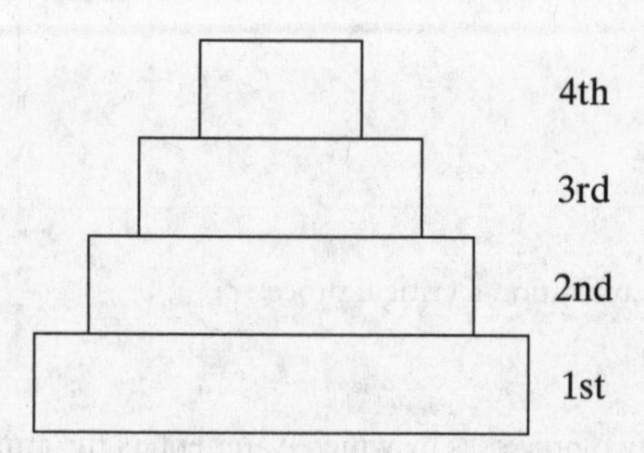

   b. Approximately how many prey animals would it take to support one predator?

3. In Chapter 15, you learned about genetic engineering. What kinds of traits would you engineer into a bacteria to improve nutrient recycling?

4. a. What proportion of the energy in a given trophic level is converted into biomass in the next trophic level?

   b. What factors account for this relatively low value of energy transfer?

5. a. What is ozone?

   b. What is the formal term for the atmospheric zone commonly called the ozone layer?

   c. Why is ozone essential to life on Earth?

   d. What is the relationship between human production of CFCs and global increases in skin cancer?

6. Explain the basis for the greenhouse effect.

7. a. Label the Earth in its orbit during winter in the Northern hemisphere.

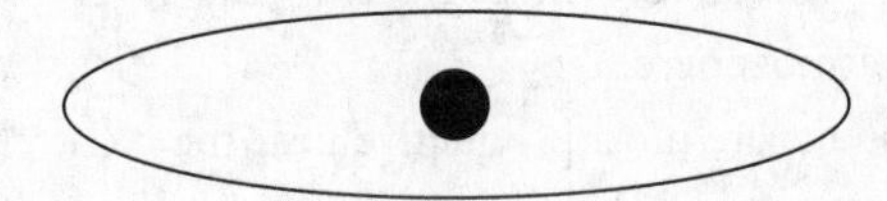

   b. Now label the Earth in its orbit during summer in the Northern hemisphere.

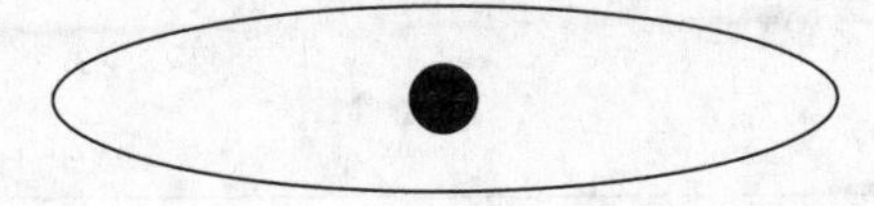

   c. Why is the Arctic cooler than Florida even during summer in the Northern hemisphere?

8. Which side of the following mountain will be relatively wet? Why?

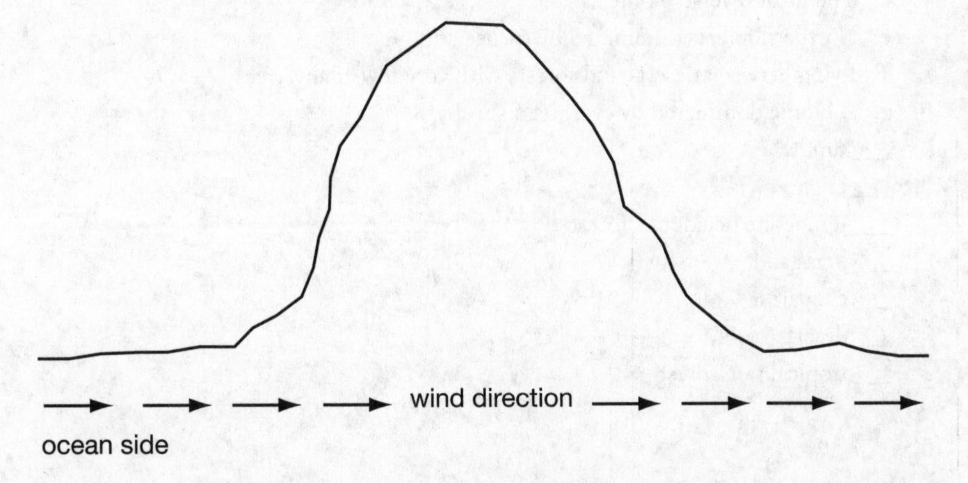

9. What is the source of the prevailing winds found at certain latitudes of the planet?

Earth's Biomes

- Climate—yearly temperature and rainfall patterns—determines the prevailing vegetation within a region that, in turn, determines the type of ecosystem present. These combinations of cold, warm, wet, and dry define six land biomes: tundra, taiga, deciduous forest, temperate grassland, desert, and tropical rain forest.
- The tundra and taiga represent the most severe climate conditions—cold and dry. Each area supports a variety of life forms, plants as well as small and large mammals; but species diversity is limited. Although there may be few different kinds of plants and animals, their numbers can be quite large.
- The deciduous forest and the temperate grasslands support a variety of animal and plant species, and each species is numerically well represented. These biomes support not only more types of plants and animals than are found in the tundra and taiga but also a greater number of each species.
- The tropical rain forest is the most productive biome, supporting the largest mass of organisms as well as the most diverse collection of species. This abundance of living matter—especially the trees—traps an enormous amount of carbon each year, primarily from the atmosphere. Thus the rain forest plays a critical role in decreasing the level of the greenhouse gas (CO_2) in the atmosphere.
- Marine ecosystems are most productive near the coast, the shallow water supporting an abundance of photosynthetic producers at the lowest trophic level. The most productive aquatic ecosystem is the coral reef, a habitat built around the accumulated carbonate skeletons of tiny sea anemones.

THINK ABOUT IT

1. Match the following descriptions of biomes with their names:

   a. Cold, low-lying, no trees
   b. Forests with beech and maple as the ecological dominants
   c. Biome with very low levels of rainfall (< 25 cm per year)
   d. The most diverse biome
   e. Northern forests dominated by spruce and pine
   f. Lying between deserts and forests with very few, if any, trees
   g. A biome dominated by evergreen shrubs

   ___ tundra
   ___ taiga
   ___ temperate deciduous forest
   ___ grassland
   ___ chaparral
   ___ deserts
   ___ tropical rain forest

2. Use the following terms to label this figure: intertidal zone, pelagic zone, open sea, coastal zone, photic zone, benthic zone.

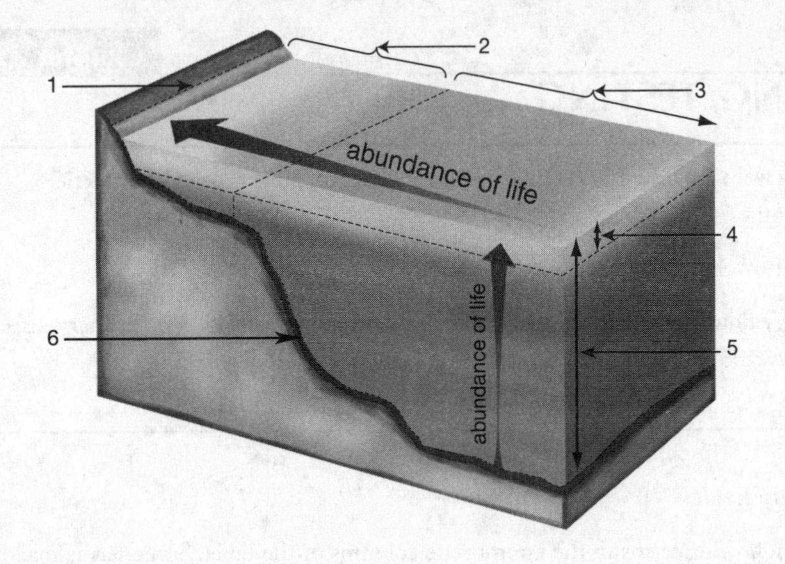

1 ______
2 ______
3 ______
4 ______
5 ______
6 ______

3. Which types of plankton are producers and which types are consumers?

4. How does nutrient enrichment result in fish kills?

Cool Fact: Water, Water Everywhere?

Freshwater occupies only about 2 percent of our planet's surface, yet it is essential to the survival of all organisms. Humans rely on freshwater sources not only to meet their drinking needs but also as a food resource, as fish taken from rivers and lakes, and as water for crops. Water also supports human interactions, particularly commerce, using waterways to transport goods. Given our absolute dependence on water, it's surprising that the current global decline in water quality and availability is not front-page news.

Water becomes scarce because of two forces—water diversion from direct human use to agricultural use (irrigation) and water contamination. Water demand for agriculture is so high that even major waterways like the Colorado River can be "drained dry" —meaning that none of the river water arrives at the sea—during the dry season. Furthermore, dams and other structures created to control water flow can cause serious habitat fragmentation, reducing food resources for humans as well as for other predator species. Water contamination denies safe supplies of drinking water to millions of people worldwide. Major water contaminants in the developing world include human fecal products and industrial wastes; in the developed world, fertilizer and pesticide runoff most often contaminate water. Current predictions suggest that by 2025, 48 percent of the world's population will reside in "water-stressed" river basins. "Water rights" are already a contentious issue in many places in the world. Is it possible to reverse the trend and manage water resources to serve both human and nonhuman needs?

The hard truth is that humans can no longer consider water a free and inexhaustible resource. Scientists can provide information that defines and maintains standards of water quality and availability, but water management comes at a cost—a cost that will need to be met by political bodies. How much is water worth? It's priceless.

TESTING, TESTING

Create a web: Aquatic Ecosystems: freshwater, ocean, photic, pelagic, limnetic, photosynthesis, coastal, littoral

Make a flowchart:

1. Energy flow: Include gross and net primary production, producers; primary, secondary, and tertiary consumers.
2. Carbon cycle

Vocabulary Review

Place the following terms in the appropriate columns of the table. Some terms may be used in more than one column.

carbon
chaparral
decomposition
fossil fuels
greenhouse effect
nitrogen
nitrogen-fixing bacteria
rain shadow
stratosphere
taiga
temperate forest
tropical rain forest
tundra

Biomes	Nutrient Cycling	Climate

PRACTICE TEST

Once you have finished studying this chapter, close your books, grab a pencil, and spend the next 15 to 20 minutes completing this practice test.

Compare and Contrast

For each of the following paired terms, write a sentence of comparison ("Both . . .") and a sentence of contrast ("However, . . .").

herbivore/detritivore

gross primary production/net primary production

estuary/wetland

ecosystem/biome

Matching

Match the following terms with their description. Each choice may be used once, more than once, or not at all.

a. Desertification
b. Biogeochemical cycling
c. Primary production
d. Eutrophication
e. Global warming

___ Change in the profile of a lake due to increased nutrient amounts
___ Formation of new desert areas
___ Carbon dioxide acts as an insulator in the Earth's atmosphere
___ Conversion of carbon dioxide to carbohydrate
___ Photosynthesis

Short Answer

1. Unlike the land beneath northern deciduous forests, the land beneath tropical rain forests is not a good candidate for sustainable farming. Explain.

2. What has changed about the carbon cycle with an increase in industrialization?

3. What are the two categories of abiotic factors that help to form ecosystems?

4. What is the greenhouse effect?

Multiple Choice

1. The island of Ecophilia is unusual because its topography allows the presence of both tropical rain forest and deciduous forest biomes. Through a series of misadventures, you have been stranded on Ecophilia for several years. To survive, you must plant a garden. Why should you choose the deciduous forest as the site for your farming attempt?
   a. because that's the way the pilgrims did it
   b. because that will leave plenty of coconut trees to use for making cool drinks
   c. because the soil receives more rain there

d. because the soil contains more nutrients there
e. because the soil receives less sunlight there

2. Elements of Earth

a. are continuously created.
b. must be recycled.
c. are continuously replenished from space.
d. can be obtained from deep within the Earth's core.
e. are not necessary for life.

3. Which of the following processes convert organic carbon (biomass) to inorganic carbon (CO_2)?

a. photosynthesis
b. respiration
c. burning of fossil fuels
d. decomposition
e. b and d

4. Which of the following organisms can fix nitrogen?

a. plants
b. fish
c. fungi
d. bacteria
e. All of the above.

5. Based on global circulation patterns, where would you expect to find deserts?

a. equator
b. 30°S
c. 30°N
d. the North Pole
e. b and c

6. Which biome has the highest species diversity?

a. chaparral
b. tundra
c. tropical rain forest
d. desert
e. temperate deciduous forest

7. Which biome is characterized by a layer of permanently frozen land?

a. tundra
b. taiga
c. boreal forest
d. cold deserts
e. None of the above.

8. Most of the Earth's water is

   a. locked up as salt water or ice.
   b. stored as carbon dioxide.
   c. radioactive.
   d. being lost to atmospheric dehydration.
   e. found in the form of steam.

9. Your friend Alfred wants to try an experiment that requires exposing seeds to the most direct sunlight possible. Where in the world should he go?

   a. to the North or South Pole
   b. to the top of Mt. McKinley
   c. to the Pacific Ocean, midway between Mexico and British Columbia
   d. to the equator
   e. sunlight strikes the Earth evenly at all points

10. How are biomes defined?

    a. by the direction of the prevailing winds
    b. by the amount of groundwater
    c. by the presence or absence of nitrogen
    d. by the type of vegetation that is found there
    e. by the type of large predator that inhabits them

11. Historically, how did nitrogen enter plants?

    a. via rocks
    b. as liquid
    c. by bacteria
    d. by snake dung
    e. All of the above.

12. Which of the following typifies desert biomes?

    a. hot temperatures
    b. low rainfall
    c. large trees
    d. sand dunes
    e. prairie grasses

13. Where is the most productive area of the ocean?

    a. near the shore
    b. at the equator
    c. low latitude
    d. open ocean
    e. low tide

BUT, WHAT'S IT ALL ABOUT?

Here's a question to help you pull together what you've learned so far using this text. If you don't remember how to attack this type of question, check out the example in Chapter 1.

A favorite topic of science fiction novels involves human efforts to colonize inhospitable environments on other celestial bodies by building self-contained biospheres. What kind of environmental challenges would the inhabitants of these artificial "earths" face in order to survive on the moon or Mars?

What do I do now?

Remember the drill—decide what the question is asking you to do, collect your evidence from this chapter (and the others you've studied) and write!

33

CHAPTER 33
Animal Behavior

Basic Concepts

- Behavioral biology seeks to determine the genetic, physiological, and environmental influences that cause animals to show specific patterns of behavior.
- Behavior can result from responses to internal cues, such as genetic predispositions, internal clocks, and hormones, or from learning actions from other members of the species.
- Social behavior, the behavior of an individual determined by interaction with other individuals of the same species, may be understood as a mechanism to increase the immediate survival of the individual as well as the long-term survival of the species' genome.

Behavioral Biology

- Behavior results from the immediate, physiological response to a stimulus, the proximate cause of the action.
- The ultimate cause of a behavior is genetic. Individuals expressing this behavior in the past have produced more offspring, which carry the genes specifying the behavior. This behavior has thus been selected for by the environment (natural selection) because it provides a reproductive advantage to the organisms that possess it.
- Behavior requires the interaction of both internal (genetic) and external (environmental) cues, as well as experience (learning).

THINK ABOUT IT

1. Which do you think has a greater influence on your daily activities, proximate or ultimate causation?

2. Once behavioral biologists have observed and described an animal's behavior, how do they determine why an animal performs this behavior?

3. How do animals choose their behavior?

Influences On Behavior

- Reflexes are immediate, physiologic responses to a stimulus that every member of a species performs the same way without conscious thought. Reflexes represent single actions, but some species also perform stereotyped patterns of behavior, such as movement, in response to a triggering stimulus. Once triggered, these action patterns are performed to completion.
- Biological rhythms, either circadian or annual, are periodic behaviors performed without an external trigger. The timing of these rhythms, however, can be adjusted by environmental signals, such as light-dark cycles, ambient temperatures, or tides.
- Hormones are signaling molecules, produced in one organ to produce an effect elsewhere in an organism. Changing hormones levels can trigger or modify behaviors.
- Learning represents the modification of behavioral responses because of knowledge gained through experience. Learning always has a genetic component, but these are behaviors that are profoundly affected by the environment, such as song acquisition in birds.

THINK ABOUT IT

1. Why do animals have such a range of ways for acquiring information?

2. How does the environment affect animal behavior?

3. How does the stereotypic behavior of action patterns differ from learned behavior?

4. a. How are circadian rhythms entrained?

   b. Would you expect that a person born blind would have the same circadian rhythms as a sighted person?

5. When Angela returned to New Jersey for the winter holidays from the University of Texas at Austin, she found it necessary to wear her winter jacket for the first two days—while she was inside the house—why?

6. Your cat Fluffy comes running to you at the sound of a whirring electric motor. What type of behavior is Fluffy showing?

7. When Fluffy was a kitten, she delighted in jumping onto the table as you sat down to eat. At first you just lifted her off the table, but the behavior continued until you began incarcerating Fluffy in her travel box, which she loathes, for 10 minutes after her tabletop excursion. After four imprisonments, Fluffy stopped interrupting your dinner—why?

8. OK, instead of just reading about behavior, go out and observe some. Then come back and apply the information discussed in the text to your observations. Here is what we want you to do:

   Go find some animal species to watch for 30 minutes (yes, you can watch people too).
   Every 5 minutes, make a note of what the organism(s) is doing.
   Fill in the chart by trying to assign the behaviors observed to one of these categories: Reflex (R), Biological Rhythm (BR), Hormonal (H), or Learning (L)
   Answer the following questions:

   a. Was any one category of behavior more prominent?

   b. Could you distinguish between proximate and ultimate causation of these behaviors?

   c. What kind of experiment would you design to determine the proximate cause of one of the behaviors you observed?

Species	Behavior	Influence

Social Behavior

- Organisms that live as part of a group may individually gain some survival and reproductive benefits, because an individual can exploit the resources (food, territory, protection) garnered by the group. But these benefits may come at a cost to the individual—namely, performing behaviors that increase group survival over individual survival.

- Dominance hierarchies represent extreme examples of individual organisms exploiting the resources gained by a social group. On the other extreme, territoriality occurs when individual organisms define territories from which they excluded members of their own species.
- Eusocial (meaning "truly social") organisms live in groups wherein the group members divide the labor among them in a highly organized, and sometimes rigid, structure. Among these cooperative groups, one individual's behavior is determined by the behavior of other members of the group.
- Altruism occurs when an individual benefits as much, or more, by increasing the reproductive success of other members of the group as it does by increasing its own reproduction. Altruism occurs most often in eusocial groups wherein the members are closely related genetically.

THINK ABOUT IT

1. Compare the social systems of cats, dogs, and humans. In the following table, fill in each box with an example of behavior that satisfies the requirements for each category listed. Overall, how would you rank each species?

	Cats	Dogs	Humans
Solitary			
Social			
Eusocial			
Type of social system			

2. Certain species of birds perform a mating behavior known as lekking. The males congregate in one area, the lek, and each one establishes a circle in which to perform his courtship dance to attract the females clustered on the outskirts of the lek. Older and stronger males usually capture the most desirable area in the center of the lek, attract more females, and prevent other males from enticing the females out of the circle. The result is that not every male who dances will attract a mate, some males will attract multiple mates, and some males will not even succeed in staking out a dance circle within the lek. Is lekking an example of territoriality or a dominance hierarchy?

3. a. My older sister marries my husband's brother, whereas my younger sister marries a close friend of the family. We all have children. When I discover that my daughter needs a bone marrow transplant, I appeal to my sisters and their families to be tested as possible donors. Who of this family group is most likely to be a suitable genetic match for my daughter?

   b. My family knows that of course, were the situations reversed, I would do the same for them. What type of behavior am I asking my family to demonstrate?

Cool Fact: Good Grief?

A compelling theory advanced to describe the ultimate cause of animal behavior suggests that an organism's suite of behaviors have been determined by natural selection because those behaviors increase the likelihood of survival of the genes that specify those behaviors. Known as the "selfish gene" theory, this model proposes that the survival of the genome outranks the survival of the individual, so that each organism becomes merely a vehicle to ensure the genome's replication and survival.

This model accounts reasonably well for some altruistic behaviors, such as kinship interactions. Some scientists believe that human kinship behavior is a prime example of the "selfish gene"; care of siblings and cousins, in addition to care of your own offspring, would increase the likelihood that more of your genes survive in the next generation. Likewise, forming a kinship bond with a mate may also enhance survival of your genes. Although spouses are genetically unrelated, mutual cooperation between them, even beyond their reproductive years, helps to increase the survival of their children—thus increasing the survival of both genetic contributions.

But can the selfish gene explain human grief at the loss of a spouse? In the cold logic of natural selection, the value of a spouse's contribution to your own genes' survival could be replaced by the behavior of a kinsperson or a new spouse—yet humans grieve. And grief can be a disabling behavior, interfering with resource gathering, care of offspring, and reproductive fitness. So why does this unproductive behavior persist?

One possible explanation is that grief is the price we pay to keep the kinship bond functional, especially the bond between unrelated individuals. Perhaps kinship and grief represent the most powerful and the most poignant example of the benefits and costs of being part of a social species.

Chapter 33

TESTING, TESTING

Create a web: Include the levels of learning and the factors that influence them.

Make a flowchart: Include the stages involved in the acquisition of knowledge.

Vocabulary Review

Place the following terms in the appropriate columns of the table. Some terms may be used in more than one column.

action patterns
altruism
circadian rhythms
classical conditioning
communication
dominance
eusocial
habituation
hierarchies
imitation
imprinting
inclusive fitness
insight
kin selection
learning
migration
operant conditioning
orientation
proximate causes
reflexes
taxis
trial-and-error
ultimate causes

Learning and Internal Influences	Social Behavior

PRACTICE TEST

Once you have finished studying this chapter, close your books, grab a pencil, and spend the next 15 to 20 minutes completing this practice test.

Compare and Contrast

For each of the following paired terms, write a sentence of comparison ("Both . . . ") and a sentence of contrast ("However, . . . ").

proximate/ultimate causation

habituation/insight

taxis/migration

altruism/kin selection

Matching

Match the following terms with their description. Each choice may be used once, more than once, or not at all.

a. Migration
b. Imitation
c. Habituation
d. Imprinting
e. Altruism

_____ Learning that takes place during a sensitive period
_____ Behavior performed at a direct cost to the actor
_____ Seasonal movement based on an internal clock
_____ Reduction in response based on repeated exposure
_____ Learning by copying another's behavior

Short Answer

1. What is the relationship between natural selection and behavioral biology?

2. Explain how song acquisition in birds could involve both learning and internal influences.

3. Describe some of the costs and benefits of being a social species.

4. How does altruism differ from reciprocal altruism?

Multiple Choice

Circle the letter that best answers the question.

1. Which of the following would probably not play a role in producing a given behavior?

   a. genes
   b. learning
   c. hormones
   d. environment
   e. All of these could play a role.

2. Which of the following is not a form of learning?

   a. imprinting
   b. reflex
   c. habituation
   d. classical conditioning
   e. operant conditioning

3. Genetically influenced movement toward or away from a stimulus is known as

   a. orientation.
   b. circadian.
   c. imprinting.
   d. taxis.
   e. buses.

4. Suppose you are asked to move home and help raise your siblings. To do so, you would have to give up on a promising relationship that could result in your only chance at marriage and children. Based on your understanding of kin selection and inclusive fitness, at least how many siblings would it take to equal your potential contribution to future generations?

   a. 1
   b. 2
   c. 3
   d. 4
   e. Infinity—no one could ever match your potential!

5. Your dog accidentally catches a paw in the refrigerator door handle while chasing a fly. As she pulls her paw free, the door opens and she discovers the repository of all those great leftovers you normally don't let her eat. Gradually, she learns to pull open the door and grab a snack whenever you leave the room. This type of learning is known as

   a. classical conditioning.
   b. operant conditioning.
   c. insight learning.
   d. altruism.
   e. habituation.

6. You and your friend Frank each adopt a duck egg being hatched in an incubator. You spend a lot of time with your duckling during the first few hours after hatching, while Frank leaves his in a box with plenty of food and water and goes to the library. The next day, your duckling tries to follow you everywhere while Frank's just wanders aimlessly. Frank can't understand the different behavior of the two until you explain. What key element did he miss in the process of imprinting his duckling?

   a. the habituation hour
   b. the dominance hierarchy
   c. the duckling adoption call
   d. the trial-and-error period
   e. the sensitive period

7. Hormones play a role in the ____________________ cause of a given behavior.

   a. proximate
   b. ultimate

c. intermediate
d. reactionary
e. genetic

8. You are returning to the island of Ecophilia to do more research. As you set up your study, you make a list of questions that you feel are essential to your understanding of *Dumbosiania* and *Mickymousiania*. Which of the following is not a question fundamental to describing the behavioral biology of these species?

   a. What do Dumbosiania and Mickymousiania do?
   b. Why do Dumbosiania and Mickymousiania perform certain behaviors?
   c. How do Dumbosiania and Mickymousiania know how to do this behavior?
   d. Where do Dumbosiania and Mickymousiania do this behavior?

9. Which of the following requires the greatest level of learning (and the least amount of genetic programming)?

   a. operant conditioning
   b. habituation
   c. insight
   d. reflex
   e. imitation

10. Which of the following would be required for migratory behavior?

    a. circadian rhythm
    b. calypso rhythm
    c. taxis
    d. annual clock
    e. imprinting

11. What is the ultimate cause of all behavior?

    a. evolution
    b. imprinting
    c. learning
    d. habituation
    e. classical conditioning

12. One day, with some time to spare on your hands, you sit by your window and watch birds forage for seeds in the yard outside. You notice that different individuals take turns acting as sentinels. That is, instead of feeding, they spend their time looking around as if in search of predators. You call your roommate over to watch as well. When your roomie asks what all the excitement is about, you explain that what you are seeing is an example of

    a. migration.
    b. habituation.
    c. altruism.
    d. territoriality.
    e. mating behavior.

BUT, WHAT'S IT ALL ABOUT?

Here's a question to help you pull together what you've learned so far using this text. If you don't remember how to attack this type of question, check out the example in Chapter 1.

Of all of the behaviors in the animal world, altruism is the most puzzling, especially when one considers the human animal. One could explain the altruistic behavior of parents for their children as merely the effect of inclusive fitness demands. However, inclusive fitness predicts that parents should stop feeling an attachment to their children once their offspring are independent, yet this is rarely the case [trust us, we're both daughters and mothers, we know]. What is the value of the continued altruism shown by parents for their children?

What do I do now?

Remember the drill—decide what the question is asking you to do, collect your evidence from this chapter (and the others you've studied) and write!

CHAPTERS 1-33

Answers

Chapter 1

A Guide to the Natural World

Sample Flow Chart for Chapter One

Atom→Molecule →Organelle →Cell →Tissue →Organ →Organism →Population

Cell ↓ Organism (bacteria)

Population ↓ Community ↓ Biosphere

Think about It

1. One example might be genetic discrimination, which is an issue that the government is confronting.
2. A unified set of principles, based on repeated observations, that explains an aspect of nature.
3. See the grouping on page 12 of this chapter for a list of properties of living and non-living things; characteristics shared by both things would not be sufficient to distinguish one from the other.
4. Evolution is the unifying thread of biology.

Vocabulary Review

Hierarchy of Life	Practice of Science	Biology
atom cell community organ	hypothesis scientific method tab variable	cell community organ evolution molecular biology hypothesis scientific method homeostasis physiology atom theory variable

PRACTICE TEST

Matching

a. (atom)
b. (theory)
c. (cell)
d. (evolution)

Compare and Contrast

organelle/organ

Both are levels of organization of life. However, organelles are found within cells, while organs are organized collections of tissues with a common function.

cell/organisms

Both are levels of organization of life. However, the cell is the building block of life, while organisms may be unicellular or multicellular individuals.

population/community

Both refer to collections of organisms. However, a population is a group of organisms in a geographic area, while a community is a collection of populations of different species in a geographic area.

atoms/molecules

Both refer to levels of organization below the level of the cell. However, atoms are the fundamental units of matter, while molecules are atoms that are chemically bonded to one another.

Short Answer

1. A hypothesis is used to design experiments that will yield facts. These facts can then be used to build a theory.
2. Falsifiable means that the hypothesis (or explanation) can be disputed. This allows scientists to discard ideas that are not supported by facts.
3. Because any single feature of a living organism may be replicated in an inanimate object.
4. In 1800 a biologist would have spent her time primarily in description of the natural world. By 1900, her emphasis would be on the manipulation of organisms through scientific experimentation.

Multiple Choice

1d, 2b, 3e, 4e, 5d, 6b, 7a, 8d

Chapter 2

Fundamental Building Blocks of Matter

Think about It

1. Your web should show these terms: protons, electrons, neutrons, mass, electrical charge, element, atomic number and isotopes.

2.

Particle Name	Location in Atom	Charge	Mass Contribution
proton	nucleus	+1	1
neutron	nucleus	0	1
electron	orbitals	−1	0

3. Atomic number is equal to the number of protons; atomic weight is the average number of protons and neutrons of the most common isotopes.
4. a, c, and e would be isotopes. b and f would be isotopes.

CHEMICAL BONDING

Think about It

1. a. proton; b. electrons
2. Water has 2 covalent bonds—the oxygen atom shares 2 electrons, one with each hydrogen atom. Oxygen is more electronegative than hydrogen and tends to pull all of the electrons toward itself, so that the oxygen side of the water molecule tends to be more negative and the hydrogen sides becomes more positive—so the covalent bonds are polar. Because unlike charges attract, the slightly more positive (hydrogen) end of one water molecule could make a hydrogen bond with the slightly more negative (oxygen) end of another water molecule.
3.

Bond	Atomic Interactions
covalent	sharing of electrons to fill the valence shell
ionic	attraction of oppositely charged ions
hydrogen	weak attractions between oppositely charged "ends" of a molecule (partially positively charged hydrogens and partially negatively charged others)

4. b, c, and d would be ions.

SOME PROPERTIES OF CHEMICAL COMPOUNDS

Think about It

1. Shape determines how a molecule interacts with its environment. For example, what other molecules it "matches."
2. a. The solvent would be the coffee (or water that made the coffee).
   b. Sugar is the solute. c. hydrogen bonds
3. It is a good solvent, it makes up the bulk of cells, life began in it, and much of life lives in it today.

Vocabulary Review

Matter	Bonds	Solutions
electron proton ion valence shell neutron isotope chemistry nucleus	ion nonpolar polar ionic covalent chemistry polar	solvent solute chemistry

PRACTICE TEST

Matching

a. (found in the nucleus, has charge and mass)
b. (found in nucleus, no charge)
c. (has charge, but no mass)
d. (has gained or lost electrons and will form bonds because of this)
e. (differs in number of neutrons from other atoms of its type)

Compare and Contrast

polar/nonpolar

Both polar and nonpolar bonds involve the joining of atoms to form compounds or molecules through the sharing of electrons. However, nonpolar bonds involve the equal sharing of electrons while polar bonds have unequally shared electrons.

inert/reactive

Both inert and reactive refer to atoms. However, reactive means that the valence shell is not filled while inert atoms have filled valence shells.

covalent/ionic

Both covalent and ionic are types of bonds. However, covalent bonds result from sharing electrons, while ionic bonds are the attractions between oppositely charged ions.

element/matter

Both terms refer to material comprising the universe. However, elements are pure forms of matter.

Short Answer

1. The different types of bonds reflect the reactivity of an atom. Depending on their configuration, atoms may gain, lose, or share electrons—thereby causing the formation of bonds.
2. Inert elements have filled outer shells and will not form bonds.
3. The atom would have fewer electrons.
4. Atoms fill their innermost shells with two electrons.

Multiple Choice

1b, 2c, 3a, 4d, 5b, 6d, 7c, 8c, 9d, 10c, 11c, 12e, 13b

BUT, WHAT'S IT ALL ABOUT?

1. Type of question already defined.

2. Evidence

Molecular Bonds	Molecular Properties
Type of bond made by atoms depends on atomic structure. Bonds made put atoms in most stable state	Solubility From chapter 1 – "like dissolves like" Molecules with polar bonds will dissolve in polar solvents
Covalent – sharing electrons between atoms, shared electrons fill up outer electron shells of participating atoms, can be polar (if electrons not shared equally—example water) or nonpolar Ionic – electrons are gained or lost, creates ions held together by electrostatic attractions, example NaCl Hydrogen bonding – always involves hydrogen (especially water hydrogens) pairing with an electronegative oxygen or nitrogen. Relatively weak bonds	Shape—2 dimensional Bonding determines how atoms line up in molecules, whether or not the molecule will have straight (H_2) or bent (H_2O) bonds Shape – 3 dimensional Determines molecule's activity, what it binds to. Example – signal molecules & receptors

3. Pull it together:

Bonds create stable structures. Atoms bond to each other, creating molecules, because making a bond puts the atoms into their most stable state. This stable state can be achieved by sharing electrons, as in a covalent bond, or by losing or gaining electrons as in an ionic bond. In both cases the type of bond affects the solubility and shape of a molecule.

The solubility of a molecule is a function of its type of bonds. Atoms can form covalent bonds by sharing electrons and filling up their outer electron shells. If the electrons are shared equally, as in methane (CH_4) the molecule will be non-polar. When the electrons are shared unequally, as in water (H_2O) the molecule will be nonpolar. Water is nonpolar because the oxygen pulls the shared electrons closer to its nucleus, gaining a slight negative charge, whereas the hydrogens, which have the electrons farther away, have a slight positive charge. Polar molecules can dissolve in polar solvents, non-polar molecules prefer non-polar solvents. Thus, the nature of the covalent bond will determine if the molecule is water-soluble (polar solvent) or not.

If one atom gives up an electron to a second atom, ions are created. Ions are charged atoms, positively charged if an electron is lost or negatively charged if an electron is gained. Because the positive and negative ions are electrostatically attracted to each other, they remain held together by an ionic bond. An example of a molecule held by an ionic bond is table salt, NaCl, and because both the sodium and the chloride are charged, salt is soluble in water.

Another property affected by bonding is the three-dimensional shape of a molecule. Both covalent and ionic bonds can produce molecules that are linear or have more elaborate shapes, such as pyramids or cubes. The nature of the bond, polar or nonpolar, also determines how individual molecules interact with each other. Water molecules, for example, can form weak bonds (hydrogen bonds) between each other because of the partial charges on the atoms. Hydrogens in larger molecules may also participate in hydrogen bonding and determine the three-dimensional shape of the molecule.

Chapter 3

Water

Think about It

1. a. Hydrogen bond formation allows linkage between water molecules and between water molecules and other types of molecules.
   b. Ice floats, allows aquatic life to survive cold temperatures.
   c. Water absorbs heat and releases it slowly, therefore acts as an insulator.
   d. Water molecules stick together (allows a water strider to walk on water).
2. Without a body of water to act as a buffer, temperatures in Tucson would vary more from day to night.
3. photophilic
4. Chemical 1 – acid, weak
   Chemical 2 – acid, strong
5. Vinegar – less than 7
   Agent X – greater than 7
   pH 4 to 3: increases $H+$ 10x
   pH 4 to 2: increases $H+$ 100x

CARBON

1. covalent, up to 4 bonds

POLYMERS FROM MONOMERS

1.

Molecule	Building Blocks	Functions 1	Function 2
carbohydrates	simple sugars	energy storage	structure
glycerides	glycerol, fatty acids	energy storage	hormones
proteins	amino acids	enzymes	structure
nucleic acids	nucleotides	genetic code	protein synthesis

2.

Carbohydrate	Complex or Simple	Function
glucose	simple	provides energy to cells
cellulose	complex	structure in plants
chitin	complex	structure in insects, fungi

3. energy and membranes
4. a. the number of hydrogens bound to carbon backbone
   b. saturated
5. a. See text figure.
   b. between the glycerol and the fatty acids
6. a. 2-carbon ring
   b. additions to the rings
   c. See text; there are many steroids.
7. Shape determines how the protein interacts with other molecules.
8. Amino acid. 1 = carboxyl group, 2 = amino group. R group confers individual properties to each type of amino acid.
9. Code for proteins, aid in protein synthesis.
10. Sugar, phosphate, nitrogen-containing base.
11. Your web is a reflection of how you relate the concepts to each other. Therefore, exact placement of the terms will vary from one student to another. If you are unsure of the relationship of any of the terms listed, go back to your text for help.

Vocabulary Review

Water	Acid/Base	Molecules
specific heat	acid	monosaccharides
hydrogen bonds	base	fats
cohesion	buffer	phospholipids
	pH	DNA
		lipids
		steroids
		oils
		lipoprotein

Think about It

Matching

a. (structural sugar in roach shells)
b. (source of insoluble fiber in the diet)
c. (energy storage in plants)
d. (energy storage in animals)
e. (organic, but not an important molecule in living cells)
a, b, c, d (made of hydrogen, oxygen, and carbon)

Compare and Contrast

hydrophobic/hydrophilic

Both refer to affinity for water. However, hydrophobic compounds are water-fearing and hydrophilic are water-loving.

monosaccharide/polysaccharide

Both are types of sugars. However, monosaccharides are single-unit sugars, while polysaccharides are multi-unit.

DNA/RNA

Both are types of nucleic acid. However, DNA stays in the nucleus while RNA can be found in both the nucleus and the cytoplasm.

lipoprotein/glycoprotein

Both are types of protein molecules. However, lipoproteins have a lipid unit and glycoproteins have a sugar unit attached.

Short Answer

1. The cohesion of water molecules, formed by the hydrogen bonds between them, holds them together creating a weak barrier known as surface tension.
2. A base by definition decreases the concentration of hydrogen ions in a solution. Adding a base would make the solution less acidic and therefore more basic.
3. Carbon is a common element with the ability to form up to four separate bonds. This property allows it to act as a "universal connector."
4. A diet high in saturated fats has been linked to increased levels of cholesterol in the blood. This in turn could lead to heart disease and/or stroke.

Multiple Choice

1c, 2c, 3e, 4b, 5a, 6e, 7c, 8e, 9b, 10d, 11b, 12a, 13d, 14a, 15c, 16a

BUT, WHAT'S IT ALL ABOUT?

1. Type of question already defined.

2. Evidence

Advantages & examples

1. Ease of construction – can assemble with one type of bond

   Examples:

   All simple sugars join to make carbohydrates the same way, splitting out water
   Amino acids join to make proteins—different proteins have different orders of amino acids.
   Nucleic acids are chains of nucleotides.
   Lipids don't follow monomer to polymer pattern, but different classes of lipids share similar structures.
   Insight #1! — If the bonds making a certain polymer are always the same that means that the cellular machinery for making the bond is always the same. It's as if everything is assembled with one size of screw, so the only tool we need is a screwdriver.
   Insight #2! – If we need only one "tool," we have an efficient assembly line that probably works pretty fast.

2. Greater diversity of molecules by arranging order of monomers

   Examples:

   Carbohydrates – linear molecules are structural, storage forms have branches
   Proteins – amino acid chain directs folding & determines shape of the protein
   From Chapter 2, we know shape also affects molecular interactions
   Nucleic acids—order of nucleotide bases contains information that directs protein synthesis
   Insight #1! – Because a molecule's activity depends on its shape (as we learned in Chapter 2), different shapes means different functions. Thus,

when we assemble monomers in different orders we should get different shapes, which means we can make a large number of molecules with many different functions using a common pool of monomers.

3. **Pull it all together**

Creating polymers from monomers offers 2 advantages to the organism: ease of molecular construction and increased diversity of molecules. Carbohydrates, proteins and nucleic acids are each polymers of monomeric units that have essentially identical structures. Because, within each class of molecules, each monomer has common features that means that the bonds joining the monomers are identical. The simple sugars that make up more complex carbohydrates are all linked by a "C-O-C" bond. Proteins are made from amino acids by joining the "C" on one end of an amino acid with the "N" of another. This means that the cell only needs one type of tool to assemble all carbohydrate polymers, one tool for all of the proteins. This implies that making polymers can happen quickly and efficiently.

In addition to easy construction, the cell can make many different molecules from the same collection of monomers just by changing the order of the monomers. In proteins, for example, by connecting the amino acids in different orders, the cell can produce many different proteins with many different shapes from the same collection of amino acids. Since the function of a protein depends upon its shape, each of these molecules can have a unique function. With the carbohydrates that are often polymers of the same sugar monomer, the cell can change the function of the polymer by where it connects the monomers. Some carbohydrates are linear molecules, like cellulose, which makes them rigid and very strong. By making branches on the chain of sugar monomers, like starch, the cell makes a polymer that is easy to store.

Chapter 4

Prokaryotic versus Eukaryotic

Think about It

1.
E dog	P bacteria	E bird	E tree
P no nucleus	E multicellular	E true nucleus	E fungus
E cell wall	E organelles	E central vacuole	E mitochondria

2. Organelles—one example of an organelle is mitochondria, the energy factories of the cell. Prokaryotes lack mitochondria.

TOUR OF THE EUKARYOTIC ANIMAL CELL

Think about It

1. a. DNA → nucleus → mRNA → nuclear pore → cytoplasm → ribosome → protein → vesicles → exocytosis
   b. Exocytosis is how the protein will leave the cell.
2. ER = production line
   Golgi = shipping department
3. a. presence or absence of bound ribosomes
   b. detoxification of chemicals/lipid synthesis
4. So they don't digest or destroy the cell.

5. mitochondria
6. Refer to Figure 4.5 to make a flow chart of the process used to make proteins.

TOUR OF THE EUKARYOTIC PLANT CELL

Think about It

1. central vacuole, chloroplast, cell wall
2. a. chloroplast
   b. mitochondria

Cell-Cell Communication

1. Plant cells have multilayer walls while animal cells rely on the fluid phospholipid bilayer.
2. Proteins

Vocabulary Review

Animal	Plant
nucleus	cell wall
plasma membrane	nucleus
mitochondria	plasma membrane
nucleolus	mitochondria
rough ER	nucleolus
lysosome	chloroplast
Golgi	rough ER
microtubule	Golgi
gap junctions	central vacuole
	microtubule
	plasmodesma

PRACTICE TEST

Matching

a. (recycling center of the cell)
b. (command center of cell)
c. (sorts and ships secretory proteins)
d. (stores material in plant cells)
e. (moves material from one location to another)

Compare and Contrast

cytosol/cytoplasm

Both are outside the nucleus. However, only cytoplasm contains the organelles.

nucleus/nucleolus

Both are not part of the cytoplasm. However, the nucleolus is found within the nucleus.

cilia/flagella

Both are microtubule-containing structures. However, cilia are shorter than flagella, and there are more cilia per cell than flagella.

smooth ER/rough ER

Both are a membrane-enclosed lumen. However, the rough ER has associated ribosomes.

Short Answer

1. These structures provide an opportunity for coordination and control in multicellular organisms.
2. Animals rely on an exo- or endoskeleton for support. This function can be replicated by a rigid cell wall and a filled central vacuole.
3. The mitochondria is like the factory power plant.
4. The proteins would probably be degraded and recycled.

Multiple Choice

1d, 2d, 3c, 4c, 5d, 6c, 7e, 8e, 9b, 10a, 11b, 12a

BUT, WHAT'S IT ALL ABOUT?

1. Have defined this for you

2. Evidence:

Class	Examples	Cellular Structures	Functions of Structures
proteins	enzymes—proteases actin	lysosomes cytoskeleton	recycling cell shape
lipids	phospholipids	membranes	compartments
carbohydrate	cellulose	plant cell walls	support
nucleic acids	DNA RNA	DNA ribosomes, mRNA	store information interpret information

3. Pull it all together

All parts of the cell are made from the 4 basic types of molecules: lipids, proteins, nucleic acids and carbohydrates. Each class of biomolecule contributes to the specialized functions of the cell's component parts. Lipids, specifically phospholipids, make up cellular membranes, so every organelle in the cell, such as mitochondria and the ER, depends on a lipid membrane for its existence. Without membranes, cells would not exist. The cell's internal skeleton is composed of proteins as are all of the enzymes that carry out cellular functions. The protein actin makes up the cytoskeletal fibers that give a cell its shape. Enzyme proteins carry out many different jobs, such as making energy inside the mitochondria and recycling biomolecules inside the lysosome. The carbohydrate cellulose makes plant cell walls rigid and strong. The nucleic acid DNA stores information in the cell nucleus. This information is copied to RNA, another nucleic acid, and used to direct synthesis of proteins on ribosomes, which are also made of RNA.

Chapter 5

Plasma Membrane—Structure and Function

Think about It

1. (1) keeping inside in, (2) keeping outside out, (3) controlled passage, (4) cell-to-cell communication
2. controlled passage of materials, in this case, chloride ions
3. a. cellular recognition
   b. cellular communication (e.g. receptors for hormones)
4. a. the glycocalyx
   b. 1. carbohydrates
      binding sites for attachment of peripheral proteins
      lubrication of cells
      adhesion of cells
   2. peripheral proteins
      structural support
5. Phospholipids and cholesterol are both nonpolar constituents of membranes, however phospholipids contain fatty acid chains and a phosphate group whereas cholesterol is a planar molecule and lacks fatty acid chains.

DIFFUSION AND OSMOSIS

1. Osmosis would likely cause the movement of water from the surroundings into the cells, causing swelling. The extreme outcome might be lysis of the cells, because animal cells have no cell walls.
2. The chloride concentration is greater inside the cells than outside, so water moves into the cells due to osmosis. This leaves less water outside the cells to dilute the mucus, so the mucus is stickier than normal.
3.

Substance	Cross solo?	Reason
Water	Yes	It is small; but because it is charged, it would move through faster with a protein.
Oxygen	Yes	Gases freely diffuse.
A large protein	No	It is too large.
A small molecule with a large charge	No	The charge will not allow transit through the lipid bilayer.
A small molecule with a large charge	Yes	Same as water

4.

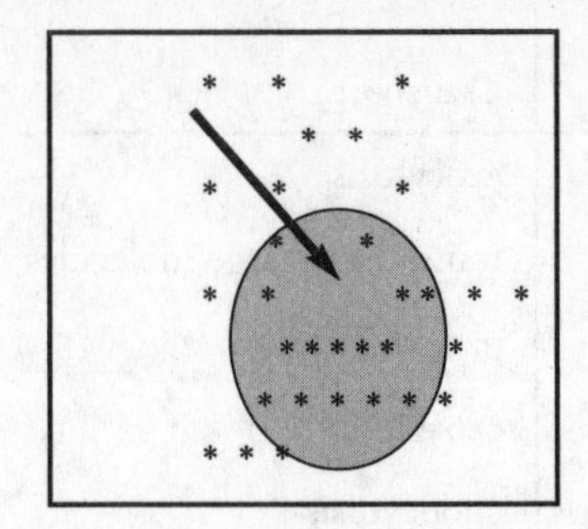

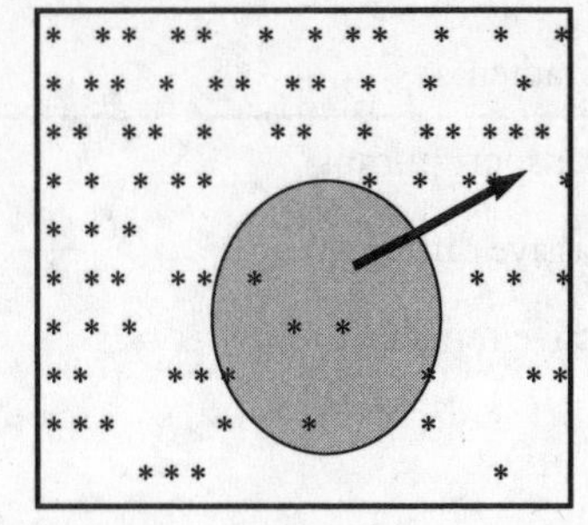

PASSIVE AND ACTIVE TRANSPORT

1.

	Simple Diffusion	Facilitated Diffusion	Active Transport
Energy Required?	No	No	Yes
Direction (with respect to a concentration gradient)	Down the concentration gradient	Down the concentration gradient	Against the concentration gradient
Membrane Proteins Required?	No	Yes	Yes
Types of Substances Transported	Very small, uncharged molecules	Small molecules, may be charged	Small molecules and ions

MOVING "BIG STUFF" IN AND OUT

Brief Review

1. a. exocytosis
   Proteins produced by the cell and destined for export
   Wastes produced by unicellular organisms
   b. phagocytosis
   Large food molecules for unicellular organisms
   Foreign invaders (e.g., bacteria) by cells of the immune system
   c. receptor-mediated endocytosis
   Specific molecules (e.g., nutrients) that bind to receptors located at specialized membrane pits
2. Draw a brief sketch of each of the above processes, with labels indicating key steps and structures.
   See Figure 5.10 in text.

Vocabulary Review

Membrane Components and Function	Gradients	Transport
plasma membranes	plasma membranes	plasma membranes
fluid mosaic model	osmosis	vesicles
proteins	diffusion gradient	transport pump

Membrane Components and Function	Gradients	Transport
fluidity	transport pumps	exocytosis
hydrophobic	passive diffusion	transport of large molecules
transport pumps	active transport	passive diffusion
phospholipid		exocytosis
communication		endocytosis
glycocalyx		phagocytosis
cholesterol		active transport
hydrophilic		

PRACTICE TEST

Matching

a. (has ends that stick out of both sides of membrane)
b. (found on a bilayer)
c. (maintains membrane fluidity and serves as a patching compound)
d. (is both "water loving" and "water fearing")
e. (serves as a binding site for proteins on the cell surface)

Compare and Contrast

phospholipids/cholesterol

Both are important components of cell membranes. However, phospholipids form the major portion of the membrane while cholesterol helps maintain its fluidity.

passive diffusion/facilitated diffusion

Both involve the movement of material down the concentration gradient across a semipermeable membrane. However, facilitated diffusion involves the use of a transport protein.

phagocytosis/pinocytosis

Both are methods of moving material into a cell. However, phagocytosis ("cell eating") requires the use of pseudopodia to engulf material.

Short Answer

1. Cell membranes are made up of hydrophobic molecules for the most part. This allows cells to maintain a barrier between the internal water-based cellular environment and its surroundings.
2. Moving down along the concentration gradient requires less energy.
3. Such materials are too large to use a transport protein as a channel into or out of the cell.
4. Water is a charged molecule; other charged molecules are unable to cross the hydrophobic lipid barrier without help, and yet water crosses freely.

Multiple Choice

1d, 2d, 3c, 4c, 5d, 6d, 7c, 8a, 9d, 10a, 11e, 12a

BUT, WHAT'S IT ALL ABOUT?

1. **What kind of question is this?**
   Answer already provided.

2. **Collect the evidence**
   Your collection of examples need not be exhaustive, just convincing, so look for diverse examples of how membranes make it possible for organisms to do what they do. Use the information in Chapter 5 to answer this question.

Features Of Living Things	Role Of Plasma Membrane(S) (PM)
Assimilate & use energy	Transport – PM helps move fuel molecules into the cell using specialized transport structures. Can concentrate molecules in cell
Respond to environment	Communication – receptor proteins and the glycocalyx transmit information about the outside world to inside the cell
Maintain constant internal environment	Structure- phospholipids create a barrier, keeping useful molecules inside and harmful ones outside. Also transport function controls entry & exit of molecules
Highly organized	Has shape determined by lipid bilayer, contains protein specific to type of cell
Made of cells	Duh, no membrane = no cell

3. **Pull it all together**

   Because plasma membranes allow living organisms to do the things they do, we can argue that the simplest definition of a living thing is something that has plasma membranes. All living things are made of cells, and cells cannot exist without a plasma membrane to separate what is inside the cell from the environment. Membranes are highly organized structures: they have a shape because of the nature of the phospholipid bilayer and this bilayer contains proteins that identify the type of cell. The plasma membrane creates a barrier, keeping useful molecules on the inside and useless and harmful ones outside, so that whatever is inside a cell will stay there until the cell decides to move it out. The receptor proteins and glycocalyx imbedded in the membrane communicate information about the outside world to the inside of the cell so that the cell can respond to its environment. The special transport proteins in the plasma membrane allow the cell to concentrate molecules on the inside that the cell can use to make energy. Thus the plasma membrane makes it possible to accomplish many of the characteristics of living organisms.

Chapter 6

Principles of Thermodynamics

Think about It

1. transform energy
2. food, sun
3. a. potential b. potential c. kinetic
4. kinetic energy plus heat
5. kinetic

HOW LIVING THINGS USE ENERGY

Think about It

1. because you spent energy and generated disordered heat
2. a. mechanical—muscle contraction
   b. transport—pumping ions across membranes
   c. synthetic—manufacture of macromolecules
3. left—exergonic; right—endergonic
4. M+++ and P−−−
5. 1—exergonic; 2—ADP; 3—P; 4—endergonic; 5—ATP

EFFICIENT ENERGY USE IN LIVING THINGS: ENZYMES

Think about It

1. to facilitate the reactions so that they occur at a reasonable rate
2. Metabolism is the sum of all chemical reactions carried out by an organism. The metabolic pathway consists of the steps required to complete a complex process.
3. a. (i) phospholipids; (ii) DNA; (iii) cellulose
   b. RNAase
4. a. b. (overall energy change of the reaction)
   a. (activation energy)
   b. (activation energy)
   c. (energy of the reactants)
   d. (energy of the products)
5. Yes, because it is not permanently changed by the reaction.
6. proteins
7. because they help certain enzymes by functioning as co-enzymes
8. a. allosteric regulation
   b. because shape is related to function
9. a. round site on top where the substrate binds
   b. the product
   c. binding of the product to the allosteric site changes the shape of the enzyme thereby changing its function (the active site closes).

Vocabulary Review

Energy	Enzymes
endergonic	co-enzyme
exergonic	vitamin
metabolism	metabolism
ATP	metabolic pathway
ADP	protein
thermodynamics	
entropy	
coupled reactions	

PRACTICE TEST

Matching

a. (what is changed by an enzyme-mediated reaction)
b. (reaction requiring energy)
c. (amount of energy required to begin a reaction)
d. (level of disorder in the universe)
e. (where enzymes and substrates meet)

Compare and Contrast

endergonic/exergonic
Both describe energy flow in a chemical reaction. However, they describe an opposite energy difference between the reactants and products of a reaction.

enzyme/co-enzyme
Both work to catalyze chemical reactions. However, a co-enzyme helps an enzyme.

potential energy/kinetic energy
Both are forms of energy. However, potential energy is stored energy, while kinetic energy is the energy of motion.

competitive inhibitor/allosteric inhibitor
Both inhibit enzyme activity. However, they do so by binding at different sites on the enzyme.

Short Answer

1. Co-enzymes help to align the substrate with the active site of an enzyme. Many vitamins act as co-enzymes in humans.
2. Energy transformations (potential to kinetic) are leaky, and much energy is lost as heat. Therefore, we must continue to supply our bodies with energy sources (food).
3. ATP molecules are units of energy that can be used for a variety of functions like movement, transport, or forming chemical bonds.
4. The energy used to actually perform some action.

Multiple Choice

1e, 2c, 3b, 4d, 5c, 6c, 7a, 8b, 9b, 10b, 11d, 12a, 13b

BUT, WHAT'S IT ALL ABOUT?

1. **What kind of question is this?**
Defined for you in the chapter.

2. **Collect the evidence**
a. First law of thermodynamics says that energy is not created or destroyed, only transformed. Transformation is the key point. Provide an example of how cells tranform energy, such as the glycogen example provided in the chapter.
b. Second law of thermodynamics says that energy transfer results in increased disorder in the universe, which is not the same thing as a system within that universe. So a system can harvest some of the energy in the universe, but not all. The energy that is dissipated into the universe increases entropy (key term), so the second law is obeyed.

3. **Pull it all together**

According to the first law of thermodynamics, energy cannot be created or destroyed only transformed. When we talk about cells making energy in the form of ATP, what we really mean is that they transform some other form of energy, like the chemical energy in the bonds holding glucose molecules together as glycogen, into the chemical energy of phosphate bonds in ATP. During this transformation, some of the energy is transformed into heat. At the end of the transformation we still have the same amount of energy, only instead of all of it being in one type of chemical bond, now it exists as another type of chemical bond plus heat.

The second law of thermodynamics states that the universe tends to become more disordered during these energy transformations; a state known as an increase in entropy. Even though cells are highly ordered systems, making these systems has converted solar energy (the energy of the universe) into chemical energy and heat. Dissipating this energy as heat means that we cannot go backwards from chemical energy to sunlight energy, so the total entropy of the universe has increased. So, even though living organisms appear to violate the laws of thermodynamics, when these energy transformations are considered on the scale of the universe, we can show that living things do obey the law.

Chapter 7

Oxidation-Reduction Reactions

Think about It

1. endergonic
2. Electrons are transferred from one molecule to another.
3. O, R, R

THE THREE STAGES OF CELLULAR RESPIRATION: GLYCOLYSIS, THE KREBS CYCLE, AND THE ELECTRON TRANSPORT CHAIN

Think about It

1. glycolysis (in cytoplasm), Krebs cycle (in mitochondria), ETC (in mitochondria)
2. NADH, pyruvic acid
3. Refer to Figure 7.5 in your textbook. The high-energy molecule that donates electrons to make NADH is glyceraldehyde-3-phosphate. The high-energy molecules that donate phosphate groups to make ATP are 1,3-diphosphoglyceric acid and 3-phosphoglyceric acid.
4. electron transport chain
5. electron transport chain
6. mitochondria
7. NADH and $FADH_2$
8. Input – acetyl coA
   Output – NADH, ATP, $FADH_2$
9. They are transferred to electron carrier molecules.

10. The flow of protons back across the membrane is used to catalyze the reaction.
11. H_2O
12. partway through glycolysis
13. glycogen

Vocabulary Review

Glycolysis	Krebs cycle	ETC
sugar splitting	acetyl CoA	ATP synthase
NADH	FAD^+	oxygen
fructose-6-phosphate	NADH	H^+pumps
pyruvic acid	mitochondria	mitochondria
ATP	citric acid cycle	ATP
ADP	ATP	H_2O
	$FADH_2$	inner membrane
	ADP	ADP

PRACTICE TEST

Matching

a. (shared by all living organisms)
b. (stage at which oxygen is required)
c. (3-carbon molecule converted to 2-carbon molecule and a molecule of CO_2 is released)
d. (takes place in inner mitochondrial compartment)
e. (a process by which an electron is lost)
f. (process by which electron is gained)

Compare and Contrast

NADH/$FADH_2$

Both are energy carriers. However, $FADH_2$ is produced only in the Krebs cycle, while NADH is produced in both glycolysis and the Krebs cycle.

mitochondria/cytoplasm

Both are sites of energy harvesting. However, glycolysis takes place in the cytoplasm, while Krebs and the ETC take place in the mitochondria.

oxidation/reduction

Both are reactions involving the transfer of electrons. However, in oxidation, electrons are lost while in reduction electrons are gained.

pyruvic acid/citric acid

Both are molecules produced during cellular respiration. However, pyruvic acid is produced by glycolysis, while citric acid is part of the Krebs cycle.

Short Answer

1. Evolution of the Krebs cycle allowed organisms to harvest energy more efficiently. This in turn would have provided the resources needed to build larger, more complex structures.

2. Coupled reactions is a term used to describe situations in which the product of one reaction becomes the substrate of the other. Redox (reduction/oxidation) reactions are prime examples of this.
3. Oxygen is not required during glycolysis.
4. CO_2 is a by-product of the breakdown of carbohydrate during acetyl CoA formation and the Krebs cycle. For each glucose molecule that enters the cellular respiration pathway, eventually six CO_2 molecules will be released.

Multiple Choice

1c, 2c, 3c, 4d, 5d, 6b, 7d, 8c, 9d, 10a, 11e, 12c

BUT, WHAT'S IT ALL ABOUT?

1. **Type of Question:**
   Answered this already
2. **Collect the Evidence:**
   a. What biomolecules have been consumed?
   Turkey – mostly protein, some fat
   Vegetables – mostly complex carbohydrates, could have some fat depending on preparation.
   Dressing, pie – mostly simple carbohydrates, and fat.
   b. What is the fate of these molecules?
   Carbohydrates – Complex carbohydrates will be broken down to simpler molecules, which will enter cellular respiration and be broken down to CO_2, water, and the energy molecules, NADH and ATP.
   Fats – These are also broken down to glycerol and fatty acids and oxidized in cellular respiration.
   Proteins – Probably broken down to amino acids and used to make new proteins
3. **Pull it all Together:**
   Our Thanksgiving feast is comprised of various biomolecules: protein and fat in the turkey, mostly carbohydrates in the dressing and vegetables, and mostly fats and simple sugars in the pie. The fats and carbohydrates can be broke down to simpler sugars and oxidized in cellular respiration to make ATP energy. The proteins will probably be broken down to amino acids, which will be used to make new proteins. Because we are usually not very active following Thanksgiving dinner, and it was a large meal, we will probably use this energy to build new molecules, such as triglycerides or proteins rather than use the ATP energy to power muscle activity.

Chapter 8

Photosynthesis and Energy

Think about It

1. No, plant cells (one type of eukaryotic cells) depend on photosynthesis for making cellulose, a structural molecule needed to make cell walls, as well as on the energy stored in glucose. Animal cells also rely on photosynthesis to provide fuel molecules.

2.

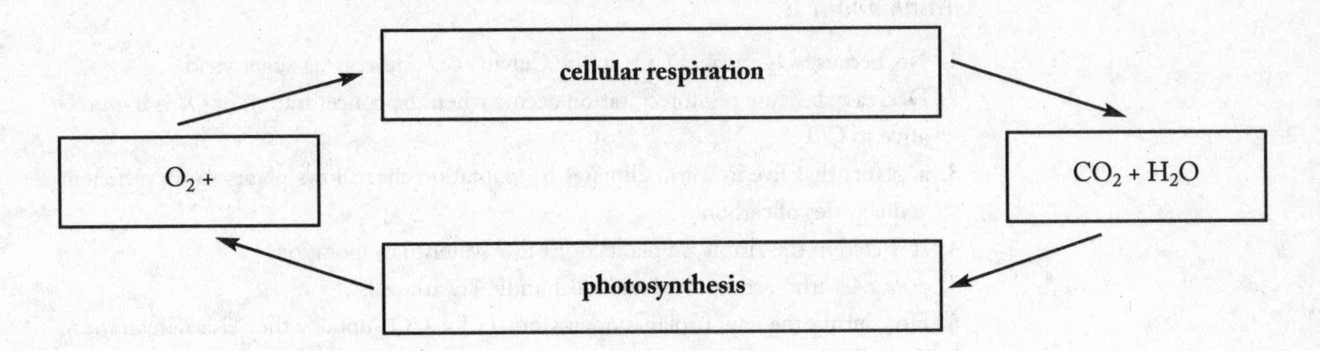

3. CO_2 in, O_2 out
4. water

TWO STAGES OF PHOTOSYNTHESIS

Think about It

1. through stomata
2. a. chloroplasts
   b. See Figure 8.3 in your textbook for illustration.
3. to convert the Sun's energy to NADPH and ATP by transferring electrons
4. sunlight
5. Primary electron acceptor → PSI → NADPH
6. a. to accept "spent" electrons
   b. the breakdown of water by photosynthesis
7. NADPH and ATP; in the chloroplast
8. photosynthesis and cellular respiration:

Similarities	Differences
Transfer of electrons through a series of carrier molecules and use the energy to make a proton concentration gradient.	Electron donor—water in photosynthesis, NADH in respiration
Use a proton gradient to provide the energy for ATP synthesis.	Products—NADPH in photosynthesis, water in respiration
Reactions occur in specialized organelles in the cell.	Photosynthesis is an endergonic process; respiration is exergonic

9. a. bringing carbon from the environment (CO_2) and "fixing" it into an organic molecule
   b. the Calvin cycle
10. an enzyme that fixes carbon, first step of the Calvin cycle
11. 3 RuBP, 6 G_3P
12. Let Figure 8.8 help you trace the path of CO_2 through photosynthesis and Figure 8.6 show you the path of O_2 through photosynthesis

ALTERNATIVE PATHWAYS OF PHOTOSYNTHESIS

Think about It

1. No, because O_2 replaces CO_2 in the Calvin cycle, decreasing sugar yield.
2. Decrease, because photorespiration occurs when the concentration of O_2 is high relative to CO_2.
3. a. plants that live in warm climates b. adaptation that allows plants to fix sufficient quantities of carbon
4. It is cold in the Arctic, so plants don't lose water to evaporation.
5. uses a 4-carbon compound and has bundle-sheath cells
6. Hot during the day; if plants open stomata for CO_2 uptake, they risk dehydration.
7. They fix CO_2 at night, so that stomata can remain closed during the day.

Vocabulary Review

Light-dependent reactions	Light-independent reactions
photosystem I	CAM
photosystem II	C4
chloroplast	stroma
NADPH	chloroplast
O_2	rubisco
light	CO_2
H_2O	RuBP
accessory pigments	NADPH
	G_3P
	glucose

PRACTICE TEST

Matching

a. (uses the energy from the Sun to build carbohydrates)
b. (releases the energy stored in food)
c. (its products are used in the Calvin cycle)
d. (the synthesis reactions of photosynthesis)
e. (when rubisco binds O_2 instead of CO_2)
f. (the currency of cellular energy)

Compare and Contrast

CAM/C4 photosynthesis

Both are adaptations to warm climates. However, CAM plants shift some activities to night, while C4 plants do not.

chloroplast/chlorophyll

Both are structures involved in photosynthesis. However, chlorophyll is a molecule, and chloroplasts are organelles.

photorespiration/photosynthesis

Both processes are related to the conversion of the Sun's energy to chemical bonds. However, photosynthesis produces carbohydrates, while photorespiration does not.

stroma/grana

Both are parts of the chloroplast. However, grana are stacks of thylakoids whereas the stroma is the fluid surrounding them.

Short Answer

1. RuBP is the first step in the Calvin cycle, and it is responsible for carbon fixation.
2. The light-dependent stage uses chlorophyll as an energy transducer.
3. NADPH provides electrons that will be used during the light-independent stage to build G_3P.
4. Carbon dioxide provides the carbons for G_3P, water provides electrons to keep the photosystems functioning, and energy is used during the endergonic process of building sugar.

Multiple Choice

1a, 2b, 3d, 4d, 5e, 6d, 7c, 8d, 9d, 10b, 11c, 12b, 13e

BUT, WHAT'S IT ALL ABOUT?

1. **Type of Question:**
   Answered this already.
2. **Collect the Evidence:**

Differences	Similarities
Photosynthesis only in plants	Both use an electron transport
Photosynthesis makes sugar, not break it down	Both make a proton gradient to make ATP
Photosynthesis makes a reduced electron carrier, cellular respiration makes an oxidized one	Both occur in specialized organelles
	Both use membrane proteins

3. **Pull it all Together:**

On the surface, photosynthesis and cellular respiration are opposing pathways. Cellular respiration breaks down sugars to make ATP energy whereas photosynthesis collects the energy in sunlight to make the ATP energy used to make sugar. Cellular respiration is an oxidative process, oxidizing sugars and producing an oxidized electron carrier. In contrast, photosynthesis is reductive, making NADPH from NADP+ and sugars from CO2 . Despite these differences, both pathways share several common features. Both occur is specialized organelles, the chloroplast for photosynthesis and the mitochondria for cellular respiration. Both pathways depend upon integral membrane proteins to create a proton gradient that is used to make ATP. In both cases, the movement of electrons is essential to making the final products. Thus, while the details differ, the pathways of photosynthesis and cellular respiration show the same pattern of organization.

Chapter 9

DNA Structure and Function

Think about It

1. a. books and libraries
   b. Because like books are part of libraries, genes are part of the genome and make it up. Both genes and genomes are in the same language (DNA).
2. a. protein
   b. by specifying an intermediate mRNA molecule

3. One-half of the parent's genome
4. DNA → mRNA in nucleus

   The mRNA is transported to cytoplasm, then to ribosome (small circle), where proteins are made.
5. because one chromosome of each pair comes from the female parent, and one comes from the male parent
6. one chromatid before DNA replication; two chromatids (one centromere) after
7. a. 23

   b. 22

   c. 1

   d. Males have two different sex chromosomes, whereas females have two copies of the same sex chromosome.

MITOSIS AND CYTOKINESIS

Think about It

1. to produce more cells (e.g., for growth)
2. a. chromosomes

   b. cytoplasm
3. G1 → S → G2 → M

   G1—B; S—C; M—A; G2—B
4. **Prophase**

   Chromosomes condense; mitotic spindle starts to form; nuclear envelope breaks down.

   Metaphase

   Chromosomes line up on the metaphase plate.

   Anaphase

   Sister chromatids separate and move to opposite sides of the cell.

   Telophase

   Chromosomes start to decondense; nuclear envelope starts to re-form.

   Cytokinesis

   Cell divides into two equivalent daughter cells.
5. anaphase
6. a. Their rigid cell wall can't be "belted" like animal cells.

   b. See your textbook for diagram. Cell plate is unique to plants; animal cells form a cleavage furrow.
7. See your textbook for diagram. Bacterial cells divide by binary fission, no obvious mitotic spindle.

Vocabulary Review

DNA	Cell division
genome	cytokinesis
DNA	centrosome
adenine	mitotic spindle
cytosine	binary fission
thymine	cleavage furrow
guanine	microtubules
chromatid	

PRACTICE TEST

Matching

a. (the splitting of one cell into two)
b. (individual packets of DNA)
c. (combination of gene and protein)
d. (the spindle organizing structure during cell division)
e. (the complete collection of genetic information)

Compare and Contrast

gene/genome
Both are made of DNA. However, a gene is one component of the genome.

chromosomes/chromatin
Both are involved in packaging DNA in the cell. However, a chromosome is one piece of DNA plus proteins, while chromatin is any DNA-protein complex.

autosome/sex chromosome
Both are normal components of a karyotype. However, autosomes are found in all human cells, whereas different sex chromosomes are found in male and female cells.

binary fission/mitosis
Both are types of cell division. However, binary fission is used by bacteria, and mitosis occurs in eukaryotic cells.

Short Answer

1. Formation of plant-cell walls begins along the metaphase plate while bacterial cell walls begin at the outer edge of the cell and move inward.
2. RNA that will result in the construction of protein or regulatory factors.
3. No, homologous chromosomes have the same genes. The Y chromosome contains the SRY sequence that determines sex.
4. Chromosomes line up on the metaphase plate and then are pulled apart, separating the sister chromatids into two groups.

Multiple Choice

1d, 2b, 3e, 4c, 5d, 6d, 7b, 8e, 9d, 10d, 11d, 12d

BUT, WHAT'S IT ALL ABOUT?

1. What type of question is this?
Compare and contrast

2. What kind of evidence do you need?
From Chapter 3, we know that proteins are very large and made up of 20 different kinds of monomeric units (amino acids). Depending on how the amino acids are arranged in the primary sequence, we can imagine an almost infinite number of different kinds of molecules.

From this chapter, we know that DNA is made up of only 4 kinds of monomers. However, it can replicate itself using each strand as a template.

3. Pull it all Together.

Proteins are assembled from 20 possible amino acids. If each amino acid represented a unit of information, then the primary sequence of a protein could carry information much like words in a sentence carry information. Protein sequences have the advantage of having almost limitless variability. The disadvantage with proteins for storing information is figuring out how to copy the sequence exactly without other proteins. Although much simpler in structure because of having only 4 possible nucleotides, DNA has a built in mechanism for making exact copies of itself. This self-replicating feature of DNA gives it a great advantage over proteins as an information storage molecule.

Chapter 10

Meiosis

Think about It

1.

Stage	Key Events
prophase I	Homologous chromosomes pair up and cross over.
metaphase I	Homologous pairs (tetrads) line up on the metaphase plate.
anaphase I	Homologous chromosomes separate.
telophase I	Nuclear envelope re-forms.
prophase II	Chromosomes condense.
metaphase II	Chromosomes line up on the metaphase plate.
anaphase II	Sister chromatids separate.
telophase II	Nuclear envelope re-forms.

2. haploid, with two chromatids
3. Four haploid cells; each chromosome has one chromatid
4. Meiosis II, because sister chromatids separate.

SIGNIFICANCE OF MEIOSIS?

Think about It

1. a. 32
   b. 16
2. Independent assortment of chromosomes and crossing over between homologous chromosomes. Siblings are rarely identical.

3.

	Alignment 1		Alignment 2	
Pair A	maternal	paternal	maternal	paternal
Pair B	maternal	paternal	paternal	maternal

REPRODUCTION: HUMAN AND OTHERWISE

Think about It

1. See Figures 10.6 and 10.7 in your textbook. Mitochondria provide energy; the tail propels the sperm.
2.

Feature	Spermatogenesis	Ogenesis
Gametes produced	4	1
Chromosomes per gamete	23	23
Arrest at any stage?	no	yes, prophase I
Gamete like a functioning cell?	no	yes
Equal cytokinesis?	yes	no

3. Sexual reproduction generates more diversity due to independent assortment, crossing over, and random fertilization of gametes.
4. Asexual reproduction allows the production of offspring that are identical to their parents; this eliminates the time required to find and negotiate with a mate.

Vocabulary Review

Meiosis	Gametogenesis
recombination	oogonium
prophase	spermatid
metaphase	polar bodies
anaphase	
telophase	
tetrad	

PRACTICE TEST

Matching

a. (meiosis II)
b. (spermatogenesis)
c. (prophase I)

d. (polar bodies)
e. (recombination)

Compare and Contrast

chromosome/chromatid

Both a chromosome and a chromatid are linear DNA sequences with associated proteins. However, a chromosome consists of one chromatid before it is replicated; after replication, there are two chromatids.

haploid/diploid

Both refer to the number of chromosomes present in a cell. However, haploid means having one set of chromosomes (the number found in gametes), while diploid refers to having two haploid sets of chromosomes (the number found in somatic cells).

meiosis/mitosis

Both are types of cell division. However, meiosis reduces the chromosome number and generates diversity, while mitosis generates identical daughter cells.

spermatocytes/oogonia

Both are cells produced during gametogenesis. However, spermatocytes are involved in male gametogenesis, while oogonia are involved in female gametogenesis.

Short Answer

1. Meiosis allows for the even reduction of chromosome number, from diploid to haploid. This in turn allows sexual reproduction to proceed.
2. Meiosis is much less common, occurring only during gametogenesis.
3. Eggs are approximately 200,000 times bigger than sperm and must contain all the material required of a functional cell.
4. Asexual offspring are exact copies of their parent.

Multiple Choice

1c, 2d, 3e, 4a, 5c, 6c, 7c, 8c, 9a, 10c, 11a, 12d, 13a

BUT, WHAT'S IT ALL ABOUT?

1. **What type of question is this?**
   Describe the effect of "A" on "B". In this case "A" is the process of meiosis and "B" is chromosomal missegregatio

2. **What kind of evidence do you need?**
   The first step is to review gamete formation in humans. If you just look at Figure 10.5, you should be struck by how similar the processes look in the male and female, the most significant difference is that in the female process meiosis I and II produce cells that are not all equal in size. But should size matter? Isn't the segregation process still the same? Segregation is the same, but what about the source of these cells and the timing? Remember the finding that missegregation happens more frequently in older humans.

3. **Pull it all Together.**
   Gamete formation in human males and females differ in two significant ways. First, gamete formation in males produces 4 equal size spermatids, but in females only a

single egg is produced, the other 3 polar body cells are degraded. If missegregation happens in either meiosis I or II, it will result in only one sperm among millions but in females could occur in the single cell that will become the oocyte. Second, the nature of the starting cells is very different. In males, sperm are produced from spermatogonia, which have the capacity to regenerate themselves. Because of this self-renewing property, males are capable of making sperm throughout their lives. The implication of this is that sperm are always "made fresh," meaning that each spermatogonium hasn't been waiting around and aging. In contrast, female gametes originate from oogonia, which are not self-renewing. Every oogonia in a female ovary has been there since embryonic life stuck in meiosis I, meaning that by the time meiosis is completed and an egg is produced, egg is at least 12-13 years old. Because missegregation happens more often as people age, it seems likely that the female's eggs are more likely to be making errors in meiosis II.

Chapter 11

Phenotype and Genotype

Think about It

1. (a) The units of heredity come in pairs.
   (b) These units retain their properties through generations.
   (c) These paired units move apart (separate) into different gametes.
2. Mendel is considered the father of genetics.
3. siblings
4. self-fertilize (self-mate)
5. The genotype is the mutation in the chloride channel gene; the phenotype is the cystic fibrosis disease (and accompanying mucus production in the lungs).

MENDEL'S STUDIES

Think about It

1. P = grandparents.
   F1 = their children.
   F2 = their grandchildren.
2. Orange is dominant. The ratio in F2 would be 3 orange:1 yellow.
3. (a) and (e) would be alleles of the same gene, as would (d) and (b).
4. Anaphase I; homologous chromosomes segregate.
5. homozygous
6. a. Purple is dominant.
   b. Plant A is homozygous, plant B is heterozygous.
7. a. Square 1:

AA	*Aa*
Aa	*aa*

   b. In Punnet square 2, the parents were *Bb x Bb*.
   c. Square 3:

	AB	***Ab***	***aB***	***ab***
AB	*AABB*	*AABb*	*AaBB*	*AaBb*
Ab	*AABb*	*AAbb*	*AaBb*	*Aabb*
aB	*AaBB*	*AaBb*	*aaBB*	*aaBb*
Ab	*AaBb*	*Aabb*	*aaBb*	*aabb*

d.

Phenotype	***A_B***	***A_bb***	***aab_***	***Aabb***
Genotypes	*AABB*	*Aabb*	*aaBB*	*Aabb*
	AABb	*Aabb*	*aaBb*	
	AaBB	*AAbb*	*aaBb*	
	AaBb			
	AABb			
	AABb			
	AaBB			
	AaBb			
	AaBb			
Totals	9	3	3	1

e. These totals are in agreement with Mendel's dihybrid phenotypic ratios. The 9:3:3:1 ratio is what is expected for the F2 progeny of a dihybrid cross involving two independently assorting genes.

8. metaphase I (alignment of homologous chromosomes on metaphase plate).

9. a.

	C	*c*
C	*CC*	*Cc*
c	*Cc*	*cc*

b. On average, one child in four (1/4 chance for each child) would be expected to be homozygous for the cystic fibrosis channel-gene mutation (*cc*).

VARIATIONS ON MENDEL'S INSIGHTS

Think about It

1.

	O	*o*
O	*OO*	*Oo*
o	*Oo*	*oo*

Phenotypic ratio—3 orange:1 white
Genotypic ratio—1 *OO:2 Oo:1 oo*
The genotypic ratio would account for a 1:2:1 phenotypic ratio (orange: pale orange: white) in cases of incomplete dominance

2. The different alleles encode different versions of protein pigments. The orange flowers have two "doses" of the orange pigment, while the white flowers have no orange pigment. Flowers with one dose of the orange pigment would have an intermediate color (such as peach) if incomplete dominance were the mechanism determining flower color.
3. AB, because the heterozygote has a unique phenotype, with both alleles expressed or "seen."
4. *AA* and *AO*—type A blood
   BB and *BO*—type B blood
   AB—type AB blood
   OO—type O blood
5. Pleiotropy, in which one allele affects multiple systems in the body.
6. because the many activities and biochemical functions of life are tied together in a complex interrelationship
7. due to a healthy lifestyle (diet, exercise), that is, environmental interactions
8. One example might be height and nutrition.

Vocabulary Review

Phenotype	Genotype	Inheritance pattern
pea color	heterozygous	incompletely dominant
trait	allele	dominant
	homozygous	recessive
		codominant

Matching

a. (different versions of the same gene)
b. (specify instructions for building proteins)
c. (physical appearance of an organism)
d. (when one gene affects multiple aspects of the phenotype)
e. (when alleles cannot mask or cover each other)

Compare and Contrast

genotype/phenotype

Both are genetic terms to describe facets of an organism. However, genotype refers to genetic composition, whereas phenotype refers to the physical appearance.

allele/gene

Both describe a DNA sequence that encodes instructions. However, the actual instructions are found in the gene, while the different variants of a gene are referred to as alleles of that gene.

Law of Segregation/Law of Independent Assortment

Both refer to the behavior of genes and chromosomes during meiosis. However, the law of segregation states that different alleles of the same gene move away from one another during meiosis, while the law of independent assortment states that the segregation of genes on different chromosomes don't influence each other.

dominant/recessive

Both describe the behavior of alleles. However, dominant alleles are always observed in the phenotype, whereas the phenotype of a recessive allele is observed only if two recessive alleles (and no dominant alleles) are present.

Short Answer

1. Traits that are codominant, incompletely dominant, or pleiotropic may have dominance ratios that appear to be non-mendelian due to the interactions between alleles and/or genes. Environmental factors may also play a role by modifying the expression of a trait.
2. The first law, segregation, is demonstrated by the separation of homologous pairs during the first round of meiosis. Independent assortment, or the mixing of alleles from both parents, is demonstrated by the orientation of the chromosomes during metaphase I.
3. A gene is a sequence of DNA that codes for a molecule of RNA.
4. The phenotype is the visible expression of genotype and takes into account such factors as dominance, codominance, incomplete dominance, and pleiotropy. However, the phenotype may be further modified by the environment. For example, traits related to living in warm climates may never be fully expressed in a cold climate.

Multiple Choice

1c, 2b, 3c, 4b, 5c, 6d, 7b, 8c, 9c, 10c, 11c, 12c, 13a

BUT, WHAT'S IT ALL ABOUT?

1. **What type of question is this?**

   Describes the effect of "A" (linked traits) on "B" (Mendel's analysis)

2. What kind of evidence do you need?

What did the ratios tell Mendel? Lack of blending in inheritance, defining traits as dominant and recessive.

Think about what the F1, F2 and F3 generations would look like for a cross of green wrinkled (yyss) with yellow smooth (YYSS) if the "Y" and "S" genes never separated, so "YS" and "ys" always go together in the gametes. How would this be different from unlinked traits?

3. Pull it all Together

Mendel's law of independent assortment arose from his observation that the peas never "lost" a trait; that a trait that seemed to disappear in the F1 generation always reappeared intact in the F2. Even in the case of linked traits, Mendel would have seen a similar pattern in the F1 generation (all yellow and smooth seeds) and the recessive traits reappearing in F2, although he probably would have been puzzled why he had a ratio of 3 yellow, smooth: 1 green, wrinkled seeds instead of 9:3:3:1 he had observed for the other 5 traits when examined in pairwise combinations. This unexpected finding might have cause him to question his law of segregation, because he would never see the trait for seed color separate from seed shape. Given how carefully Mendel kept records, and his proven insight into the mechanism of inheritance despite knowing nothing about genes, I believe he would have concluded that seed color and shape were a single trait and not two.

Chapter 12

Sex-Linked Inheritance in Humans

1. Because males have only one X chromosome; thus, they don't have a second X chromosome to provide a second "healthy" allele.
2. a. Yes, it could be genetic because it is found in many family members.
   b. Yes, it could be sex-linked because more males than females are affected.

MALFUNCTIONING CHROMOSOMES

1. Same proportion of heterozygous offspring in either case (1/2):

	S	*s*
S	*SS*	*Ss*
s	*Ss*	*ss*

	S	*s*
S	*SS*	*Ss*
S	*SS*	*Ss*

2. Yes, the child could have inherited Marfan syndrome from an affected parent.
3. miscarriage
4. P/A (diploid) P/A (haploid) P (polyploidy)
5. Chromosome numbers 45 and 47 are aneuploid in humans.

6. See Figure 12.5 in your textbook.
7. Down syndrome.
8. Turner syndrome, Klinefelter syndrome.

STRUCTURAL ABERRATIONS IN CHROMOSOMES

1. a. deletion of A—B C D E
   b. inversion between B and D—A D C B E
   c. duplication of C—A B C C D E

Vocabulary Review

Sex Linked	Autosomal
Klinefelter	Huntington
Turner	cri-du-chat
red-green color blindness	sickle-cell anemia
hemophilia	Down syndrome
aneuploidy	aneuploidy

PRACTICE TEST

Matching

a. (a segment of the sequence is turned over and reinserted)
b. (exchange of material between non-homologous chromosomes)
c. (a portion of the message is completely lost)
d. (certain sequences are copied more often than others)
e. (chromosomes fail to properly sort during meiosis)

Compare and Contrast

aneuploidy/polyploidy

Both refer to abnormal chromosome numbers. However, aneuploidy is an abnormality of one or a few chromosomes, while polyploidy refers to having an extra haploid set of chromosomes.

inversion/translocation

Both are structural abnormalities of chromosomes. However, an inversion is a flipping of a sequence within a chromosome, while a translocation is an exchange between non-homologous chromosomes.

Turner/Klinefelter

Both are sex-chromosome abnormalities. However, Turner syndrome patients have only one X chromosome, while individuals with Klinefelter syndrome have two X chromosomes and a Y chromosome.

sex-linked recessive/autosomal recessive

Both are inheritance patterns for genetic disorders. However, in sex-linked recessive inheritance, the altered gene is on the X chromosome, while in autosomal recessive the altered gene is an autosome (chromosomes 1–22 in humans).

Short Answer

1. Because male humans have only one X chromosome, they are more likely to express traits that are recessive in females who have two Xs.
2. First, the disease would become much more common in the population. Then, depending on its age of onset, it could end up killing all the heterozygotes as well as homozygotes, thereby disappearing altogether from the population.
3. A translocation is the movement of genetic material from one chromosome to another, while non-disjunction is a failure of chromosomes to sort properly during meiosis, resulting in aneuploidy (an unusual chromosome number).
4. Polyploidy is very rare; in fact, in humans there are no known cases of individuals surviving much beyond birth as a polyploid.

Multiple Choice

1a, 2a, 3c, 4c, 5d, 6c, 7e, 8a, 9d, 10d, 11c, 12b

BUT, WHAT'S IT ALL ABOUT?

1. What type of question is this?

Defend a position – in this case the position has been defined by the observation that mature twins are not indistinguishable

4. What kind of evidence do you need?

Think about what you've learned in this chapter about inheritance to make some predictions about the genotype of these twins. Then think about the definitions of genotype and phenotype that you learned in Chapter 11. Is genotype the only factor that can affect phenotype?

5. Pull it all Together

Because identical twins result from a single egg giving rise to 2 fetuses, this is a case of natural cloning. Both children will have identical genotypes, which we could predict to have identical phenotypes. However, identical twins provide compelling evidence that the environment ("nurture") may have significant effects upon appearance and behavior ("nature"). Just as plants exposed to different growth conditions of altitude, water and food can grow to be different sizes, so too can differences in the environment affect the development of people. Even twins growing up in the same household cannot have exactly the same number and kind of experiences; at some point they will have unique experiences, such as being placed in different classrooms in school, that may affect future behavior. Because of their shared genes, they will probably always look alike, share common mannerisms, and have similar preferences for activities, but differences in how they are treated by family, teachers and friends will shape their personalities differently.

Watson and Crick

Think about It

1. by interpreting x-ray diffraction patterns and knowledge of the composition of DNA
2. Rosalind Franklin and Maurice Wilkins, who provided the x-ray diffraction patterns

THE DOUBLE HELIX

1. T A G G C T A G
2. polymerases and ligases
3. Replication is not perfectly accurate. However, the editing activity of DNA polymerase helps to correct mistakes.
4. sugar + phosphate + base → nucleotide → DNA → gene → chromosome

MUTATIONS

1. yes
2. skin cancer
3. No; because the gametes did not have the mutation, only the skin cell and its daughter cells will have the mutation. It cannot be passed on through the germ line.
4. Mutations are not always harmful; they are necessary for evolutionary adaptation.

Vocabulary Review

DNA replication	Mutation
DNA polymerase	point mutation
DNA ligase	cancer
A-T pair	DNA polymerase
G-C pair	A-C pair
nucleotide sequence	G-T pair

PRACTICE TEST

Matching

a. (the sugar portion of a nucleotide)
b. (one of the nitrogen-containing bases)
c. (process of creating "new" chromosomes)
d. (a change in the gene sequence)
e. (the assembly and error-checking molecule)

Compare and Contrast

DNA polymerase/DNA ligase

Both are enzymes involved in DNA replication. However, DNA polymerase catalyzes the synthesis of a new DNA strand, while DNA ligase "strengthens" the bonds.

somatic cell mutation/germ-line mutation

Both are permanent changes to the DNA sequence. However, a somatic cell mutation cannot be passed on to offspring, while a germ-line mutation can.

phosphate/base

Both are parts of a nucleotide. However, the phosphate group is found in the DNA backbone, and the base is found in the center part of the helix.

cancer/genetic variability

Both are processes arising from mutations. However, cancer results in uncontrolled cell division, while genetic variability can be a source of evolutionary adaptability.

Short Answer

1. The three components are phosphate group, deoxyribose sugar, and bases.
2. The sides of the structure are made up of sugars and phosphates connecting consecutive nucleotides.
3. To bond nucleotides together and correct mistakes in their sequence.
4. By increasing the rate of DNA mutation, carcinogens can cause the loss of cellular controls, which leads to cancer.

Multiple Choice

1b, 2e, 3e, 4d, 5b, 6e, 7d, 8c, 9c, 10b, 11c, 12d, 13a

BUT, WHAT'S IT ALL ABOUT?

1. **What type of question is this?**

   Compare and contrast – how are mutations in gametes different from those that arise in somatic cells?

2. **What kind of evidence do you need?**

   From this chapter, we know that 1 mutation is usually not enough to produce a phenotype like cancer, rather we need several mutations. Also, from Chapters 11 & 12, we've learned about pleiotropy and the types of genetic aberrations that produce disease.

   DNA replication – extremely accurate, so spontaneous mutations very rare. Mitosis – also extremely accurate, so a rare mutation will be propagated faithfully in somatic cells.

   Gametogenesis – Meiosis, so it is a reductive process, meaning that if a mutation happens during DNA replication in meiosis I, then only one of the 4 gametes produced in meiosis II will carry the mutation.

   Mutations – Can happen spontaneously during DNA replication, but can also be induced by outside agents, such as UV light, cigarette smoke, toxins, or viruses.

3. **Pull it all Together**

   Changes in phenotype caused by changes in the DNA generally require more than a single mutation event, which decreases the odds that we ever see the result of a mutation. Heritable mutations are those that happen during the process of meiosis that creates the gametes. Presuming that a mutation happens during meiosis I and is not repaired, only one of the 4 gametes produced would carry the mutant gene. Assuming the mutation is recessive, this mutant gamete would have to fuse with another gamete carrying a mutation in the same gene for a defective phenotype to be expressed in the organism. Considering the huge number of gametes that organisms produce, the rarity of spontaneous mutations, and the low probability that two mutations in the same gene would end up in the same fertilized egg explains why heritable mutations are so rare.

Mutations in somatic cells, while still rare, are passed along to each daughter cell during mitosis. During subsequent divisions, the mutation will be preserved in the DNA so there is no way to "lose" the mutation. Also, if these mutant cells accumulate other mutations over time then the odds that we will see a change in phenotype will increase. So even though spontaneous mutations are rare, the accuracy of DNA replication in mitosis and lifespan of somatic cells means that it is more likely to develop cancer than a new heritable mutation.

Transcription

1. Transcription occurs in the nucleus, and RNA is its major product.
2. a. AUG CCA UGG UAA CGC GUU
   b. Hydrogen bonding between complementary bases directs the tRNA to bring in the amino acid.
   c. messenger RNA (mRNA), or transcript
3. RNA polymerase

Translation

1. Translation occurs in the cytoplasm (on ribosomes).
2. mRNA encodes the information concerning which amino acids to add, tRNA brings each amino acid to the growing polypeptide chain.
3. a. six codons
   b. transcription
   c. mRNA
4. Redundancy means there is more than one codon that specifies a particular amino acid.
5. because the genetic code is universal (the same in all living organisms)
6. exon 1-intron 1-exon 2-intron 2-exon 3 → exon 1-exon 2-exon 3
7. A ribosome is made up of rRNA and proteins.
8. See Figure 14.9 in your textbook for the shape of a tRNA molecule. The anticodon should be UAC, and methionine should be the attached amino acid.
9. due to a termination codon entering the A site of the ribosome

Genetic Regulation

1. A promoter provides a binding site for RNA polymerase to start transcription.
2. No; in bacteria, an operon consists of many genes transcribed from one promoter.
3. a. See Figure 14.12 in your textbook for an illustration of the lac operon.
   b. The repressor is made of protein.
   c. It would always make the lactose utilization proteins, even in the absence of lactose. This would be wasteful.

d.

	Lactose Present	Lactose Absent
Repressor Activity	Inactive (can't bind operator)	Active (binds operator)

4. Humans (like all eukaryotes) do not have operons.
5. No; some genes encode rRNA and tRNA molecules.
6. Yes, because a gene includes the protein-coding segments and the regulatory sequences.

Vocabulary

Transcription	Translation
RNA polymerase	tRNA
tRNA (a product of transcription)	mRNA
mRNA	amino acid
DNA	anticodon
gene	codon
intron	rRNA
exon	polypeptide
operon	
rRNA (also a product of transcription)	

PRACTICE TEST

Matching

a. (forms part of the ribosome)
b. (substance that induces transcription of the lac operon)
c. (part of tRNA)
d. (mRNA as it leaves the ribosome)
e. (noncoding segment of DNA)

Compare and Contrast

transcription/translation

Both are involved in expressing a gene. However, transcription converts the information from DNA to mRNA, while translation produces the protein product.

tRNA/mRNA

Both are types of RNA produced by transcription. However, tRNA is involved in bringing amino acids to the growing polypeptide, and mRNA encodes the order of amino acids.

operator/repressor

Both are involved in operon regulation in bacteria. However, the operator is a DNA sequence to which the protein repressor binds.

intron/exon

Both are parts of eukaryotic genes. However, introns do not code for protein, while exons do.

Short Answer

1. Through their work with bacteria, Jacob and Monod were the first to demonstrate a mechanism by which protein synthesis (aka gene expression) could be controlled.
2. Proteins can be quite long, which means that there are many, many different possible combinations of amino acids.
3. The anticodon is a sequence located on the transfer RNA that matches the amino-acid-defining codon located on the mRNA. The anticodon-codon binding allows the proper matching of amino acid and nucleotide sequence.
4.

Molecule	Location	Structure	Function
DNA	Nucleus	Deoxyribose, phosphate, ATGC	Information storage (genes)
RNA	Nucleus and cytoplasm	Ribose, phosphate, AUGC	Messenger, transfer, ribosomal functions

Multiple Choice

1b, 2c, 3d, 4e, 5a, 6e, 7c, 8d, 9b, 10a, 11b, 12c

BUT, WHAT'S IT ALL ABOUT?

1. **Type of Question**
   Variation on defend a position – asking to explain an observation.
2. **What kind of evidence do you need?**
   Think DNA structure, particularly base composition.
3. **Sample answer**
   The sequence of nucleotides in the mRNA that an organism needs to translate into protein depends on the sequence of nucleotides in the DNA. It is possible that different organisms might have a bias in their DNA so that there are more AT nucleotide pairs than GC pairs. More AT pairs might increase the frequency of uracil in the mRNA, so there would be slightly more triplet pairs that include a U, biasing the code.

Some Tools of Biotechnology

Think about It

1. Human pituitary glands naturally make HGH; E. coli bacteria synthesize it in biotechnology.
2. The universal genetic code allows human proteins to be made in bacteria (see Chapter 14).
3. Restriction enzymes cut both strands of a DNA molecule at specific sites.
4. vectors
5. a. Cut the human gene out of its chromosome with a restriction enzyme.
   b. Cut the plasmid vector with the same restriction enzyme.
   c. Recombine the gene and plasmid vector.
   d. Transform the recombinant molecule into bacteria.
   e. Replicate and express the recombinant molecule in bacteria.
6. Cloning is making an exact copy.
   To clone your best friend:
   a. Remove a cell from your friend.
   b. Fuse that cell with an egg.
   c. Let the embryo develop.
   d. Implant the embryo in a surrogate mother.
   e. Wait for your clone to be born.
7. Cloned organisms arise by mitosis.
8. PCR has allowed the analysis of minute quantities of evidence and has facilitated DNA "fingerprinting."
9. PCR amplifies (makes many copies of) DNA sequences, not proteins.
10. Positive pole → 500–1000–2000 → Negative pole

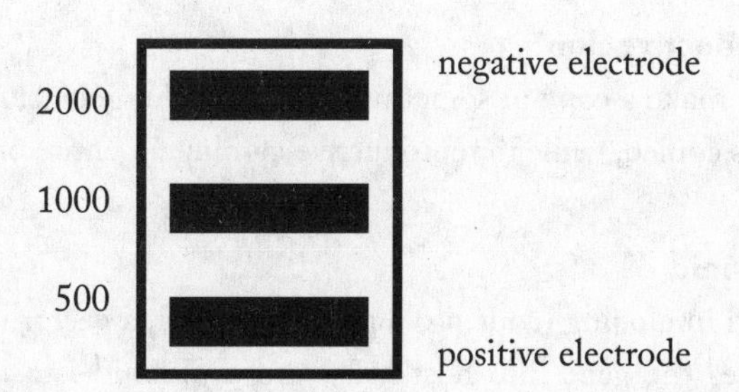

11. You could match the sequence from a suspect with crime scene evidence. In basic research, knowledge of a DNA sequence can provide information about what kind of a protein a gene may encode. This can be important to understanding disease processes.

THE HUMAN GENOME PROJECT

Think about It

1. No, in many cases the sequence of a gene does not provide information about its function.
2. Comparing the sequence of the defective gene with other human genes may provide information about the defective gene and underlying disease process.

BIOTECHNOLOGY AND ETHICS

1. The patient would no longer need injections, because their own cells would be making the clotting factor.
2. by determining which scrolls are on which animal skins and then piecing individual skins together
3. This is a good question to discuss with your friends and classmates.
4. the potential for abuse of genetic information (such as inherited disorders, susceptibility to disease)

Vocabulary Review

Cloning Tools	Applications
vector bacteriophage plasmid restriction enzymes	gene therapy biopharmaceutical forensics Human Genome Project

PRACTICE TEST

Matching

a. (molecular scissors)
b. (a gene sequence from more than one origin)
c. (a rapid way to amplify DNA in the laboratory)
d. (a molecule used to carry foreign genes into bacteria)
e. (a way to separate DNA fragments based on their size)

Compare and Contrast

gene cloning/reproductive cloning

Both are ways to make a copy of something. However, in gene cloning, a particular DNA sequence is copied, while in reproductive cloning an entire organism is exactly duplicated.

cloning vector/plasmid

Both are involved in cloning (copying) a gene. However, a vector is a type of molecule, which carries the gene into host cells, and a plasmid is a particular type of cloning vector.

polymerase chain reaction/DNA sequencing

Both are methods for studying DNA in the laboratory. However, PCR makes many copies of a DNA molecule, while DNA sequencing determines the order of nucleotides within a given DNA molecule.

proteomics/genomics

Both are ways of studying molecular biology. However, proteomics looks at the total protein output from a genome, while genomics compares genomes across species.

Short Answer

1. Xenotransplantation uses tissue (organs) from one species and transplants it into another.

2. Electrophoresis gels separate molecules by charge and size.
3. The banding patterns on gels created by cutting DNA with restriction enzymes each have a particular distribution within the population. By comparing a number of these, it is possible to identify an individual with a likelihood of 1/10,000,000 or less of making a mistake.
4. The moral and ethical issues surrounding such areas as xenotransplantation, human cloning, and genetically modified food require that time and money be spent to understand the impact of these techniques on our society.

Multiple Choice

1c, 2d, 3c, 4a, 5c, 6a, 7d, 8a, 9d, 10e, 11d, 12b

BUT, WHAT'S IT ALL ABOUT?

1. **Type of Question**

   Variation on effect of "A" on "B", where "A" is forensic analysis on "B" problem of identifying victims of a mass disaster.

2. **What kind of evidence do you need?**

   You are a forensic technician, suggesting that you will use molecular identification methods rather than physical ones, especially since there isn't much useful physical evidence left. What techniques would you use? Why? How are you going to be sure that you are correct in your identifications?

3. **Sample Answer**

   The bombing of the World Trade Center left few of its victims intact enough so that traditional methods of identification (fingerprints, dental records) were useful; the method of choice would be forensic DNA typing. The first task would be to collect blood from the parents, siblings or children who lost a loved one in the disaster because DNA typing requires us to compare DNA patterns from a sample of unknown identity to the patterns in identified samples. From each sample of human remains collected on the site, a sample of DNA would have to be amplified using PCR, cut using restriction enzymes, fragments separated by gel electrophoresis, and then compared to the patterns generated by the DNA of family members. Every sample taken from the site must be compared against every sample taken from every family who believes they lost someone; thousands of unknowns must be compared against thousands of knowns at multiple DNA sequences in order to be certain of the final identification.

Chapter 16

Common Descent with Modification

Think about It

1. Living things are modified in successive generations, which can result in the formation of new species
2. Organisms that are the most fit will pass on traits to the next generation with the highest frequency.
3. one

4. a. naturalist, making observations of the living world
   b. He had studied theology, and earned a divinity degree.
5. a. 1809 (before the U.S.-U.K. war)
   b. 1831–1836 (before the cell theory was published)
   c. 1859 (between the U.S. Civil War and Mendel's publication)
6. that organisms passed on acquired characteristics (such as webbing on a duck's feet)
7. that there had been a series of extinctions, with different fossils present after the catastrophes
8. As suggested by Malthus, there is a struggle to survive with limited resources, so that favorable variations survive and unfavorable ones do not.
9. They all represented different species related to an ancestral mainland species.
10. The idea that natural resources are limiting, so that organisms must compete with one another to access resources for survival.
11. natural selection
12. This is a personal question, although many scientists are asked to review the work of peers in an ethical and fair manner (without divulging the information or using it for their own work).

MULTIPLE LINES OF EVIDENCE FOR THE THEORY OF EVOLUTION

Think about It

1. They evolved from a common ancestor, which also possessed a notochord.
2. Terrestrial species must have evolved from an ancestral aquatic species.
3. Natural selection as a mechanism was less accepted than evolution as a process.
4. alleles of genes
5. It provided evidence from a variety of disciplines.
6. Taxonomists, who characterize species and their distributions; mathematical geneticists, who provide and test models of evolution; and paleontologists, who study the fossil record.
7. during the 1940s
8. Fossils are dated by their placement and by radiometric dating.
9. Homologous features are similar structures. The similarity between homologous structures is due to inheritance of the structure from a common ancestor.
10. A and C are more closely related because they have fewer differences in their nucleotide sequence.
11.

Radiometric Dating	Radioactive Decay Of Elements In Fossils
Placement of fossils	Similar fossils are found in similar sedimentary layers in different geographic areas, with the more advanced fossils found in more recent sediments.
Comparative morphology and embryology	Homologous structures during development and in adults of different species
Molecular biology	Similar nucleotide sequences in genes from different species

12. a. They share similar features, lifestyles, and ancestors.
    b. As evidence, the scientists used dentition (tooth structure and placement), molecular biology, comparative embryology, and the fossil record.

Vocabulary Review

Evolutionary Mechanisms	Evidence for Evolution
natural selection	fossil record
limited resources	comparative embryology
common descent with modification	homologous features

PRACTICE TEST

Matching

a. (Lamarck)
b. (Lyell)
c. (Malthus)
d. (Wallace)
e. (Darwin)

Compare and Contrast

paleontology/taxonomy
Both are disciplines that provide evidence for evolution. However, paleontology is the study of (often) extinct animals in the fossil record, and taxonomy is the study of living animals and their distribution.

natural selection/fitness
Both are related to the mechanism of evolution. Natural selection is the mechanism by which evolution occurs, and it depends on the fitness of individual organisms (how well an organism interacts with, and survives in, its environment).

embryology/homologous structures
Both provide evidence for evolution. However, embryology is the study of how animals develop, and homologous structures are similar due to inheritance from a common ancestor.

evolution/natural selection
Both describe how organisms change over time into new species. However, evolution is the process of this change, while natural selection is the mechanism by which the change occurs.

Short Answer

1. All species on Earth have descended from other species, and a single, common ancestor lies at the base of the evolutionary tree.
2. Alfred Russel Wallace
3. Yes, the work of Endler on the coloration of guppy fish in response to predation (and sexual selection) demonstrated evolution in action.
4. On the voyages of the HMS *Beagle*.

Multiple Choice

1c, 2a, 3c, 4c, 5d, 6e, 7e, 8c, 9a, 10d, 11a, 12c

BUT, WHAT'S IT ALL ABOUT?

1. **Type of Question**

    Variation on defend a position – asks you to apply your knowledge to defend a prediction.

2. **Evidence Needed?**

    Genetics – Think about allelic variation, meaning that any given gene has enough allelic variants such that phenotypes describe a range of values (like human height) and not just discrete points.

    Natural Selection – An individuals genotype allows it to survive and produce offspring under a range of environmental conditions, but any given organism may produce more offspring than other individuals in the population if it is more successful at growing and breeding under the existing environmental conditions. Common descent with modification – Offspring can only inherit the genes present in the parents, but natural selection favors the survival of some individuals over others. Over time, the favored traits predominate until eventually the offspring may look very different from the ancestral organism.

3. **Sample Answer**

    Within any population of organisms exists variation in the genotypes that produce a range of possible phenotypes. It is this range of phenotypes that are acted upon by forces in the environment, with the most favorable forms selected for survival and increased reproduction. As long as the environment remains stable, these phenotypes will be fixed in the population. Global warming is an example of a change in environmental conditions that would drive selection of more heat tolerant organisms. I would predict that if the regions of tropical climate increased on the Earth, then the animals among the current populations that could best adjust to this change would survive. Rabbits with thin fur or tough skins might be more successful in a warmer climate – these phenotypes would eventually replace the current phenotype we know for rabbits.

Evolution and Populations

Think about It

1. population
2. the gene pool (all of the alleles found in a population)
3. species
4. the change of the frequencies of alleles in a population over time
5. Natural selection; organisms that fit their environment better pass on those alleles at higher frequencies.
6. a. microevolution
   b. macroevolution
   c. macroevolution
   d. microevolution

FIVE AGENTS OF MICROEVOLUTION

Think about It

1.

Factor	How Effect Is Exerted	Is Effect Adaptive?
mutation	Change in the alleles in the population	mostly not adaptive
gene flow	New alleles arrive from other populations	random
genetic drift	Change in allele frequencies resulting from chance events	random
nonrandom mating	Different individuals pass on alleles to the next generation at different rates	random
natural selection	Organisms that fit the environment best pass on more alleles to future generations	adaptive

2. Drift is more likely to dramatically influence allele frequencies in small populations, which are more easily perturbed by chance events.
3. a. bottleneck effect
   Example: The near extinction of a species followed by recovery of the population from a limited number of individuals.
   b. founder effect
   Example: Many groups of people have arisen from a small number of founders who emigrated from their homelands, often for religious reasons.
4. DNA first, then in the mRNA transcribed from that DNA
5. The arrival of immigrants from many countries has influenced the population of North America. The high numbers of people of northern European descent have

contributed to the frequency of cystic fibrosis in North America, as well as to our cuisine, music, writing, and so on.

6. epidemic B; pioneers F
7. The relative rate at which an organism passes on its genes to future generations. The more alleles an organism passes on compared to other organisms, the fitter is the organism.
8. No, because fitness reflects the fit with the current environment, which can change. Environmental changes will change the "definition" of fitness characteristics.
9. They are both fit in their respective environments. A field mouse would not be fit in the eagle's environment, and the eagle would not be fit in the field mouse's environment.
10. The fittest individuals will leave proportionately more offspring and thus transmit more alleles to the next generation.

THREE MODES OF NATURAL SELECTION

Think about It

1. stabilizing selection
2. a. 10 dimes
   b. 100 pennies
   c. 2 quarters and 50 pennies
3. a. D
   b. S
   c. R
   d. D

Vocabulary Review

Agents of Microevolution	Targets of Natural Selection
founder effect	fitness
gene flow	allele
genetic drift	population
migration	gene pool
natural selection	genotype
nonrandom mating	
bottleneck effect	

Compare and Contrast

microevolution/macroevolution

Both are evolutionary processes. However, microevolution is a change in the allele frequency in a population, whereas macroevolution is a dramatic change in a population eventually leading to a new species.

gene flow/genetic drift

Both are agents of microevolution. However, gene flow is the movement of alleles between populations, whereas genetic drift is the chance change in allele frequency (e.g., through the reduction in population size in a bottleneck effect).

gene pool/population

Both refer to groups of individuals. However, the gene pool is the totality of alleles present in a population, whereas a population is a group of individuals living together and reproducing.

fitness/adaptation

Both refer to how well organisms interact with their environments. However, fitness refers to the number of offspring individuals leave, whereas adaptation is the change in organisms over time to better fit their environment.

Matching

a. (directional selection)
b. (genetic drift)
c. (gene pool)
d. (microevolution)
e. (fitness)

Short Answer

1. Only natural selection consistently pushes populations toward greater adaptation to their environments, whereas mutation may be random.
2. The bottleneck effect describes a reduction in the diversity of a gene pool because of a decrease in population size and implies that the population has remained in its original area. The founder effect is the result of a small segment of a population colonizing a new area.
3. The individual with the greater reproductive success (larger number of surviving offspring) is more fit.
4. Sexual selection is a form of nonrandom mating. If mating is not random, then individuals will have different levels of reproductive success, which will in turn change gene-pool frequencies.

Multiple Choice

1d, 2b, 3a, 4c, 5d, 6d, 7a, 8c, 9c, 10c, 11c, 12b

BUT, WHAT'S IT ALL ABOUT?

1. **Type of Question**

   Variation on defend a position – asks you to apply your knowledge to defend a prediction.
2. **Evidence Needed**

   Mutations – Mutations can be major events, such as when chromosomal structure changes (chapter —inversions, deletions, rearrangements), or subtler, such as point mutations. In either case, they are generally neutral or bad. But, if a mutation occurs in a gene that is pleiotropic (Chapter 11), the mutation could affect many genes at the same time.

   Natural selection – Mutations are the fuel that drives natural selection (this chapter). A mutation, or collection of them, that changes an organism's phenotype such that the change confers some advantage would allow that organism to survive better and produce more offspring, increasing the frequency of the phenotype.

3. **Sample Answer**

   Looking at any complex structure in a living organism, it is hard to imagine how accumulating mutations could produce an eye. However, genes code for molecules that execute functions, like enzymes, as well as for molecules that control how, when and where these functions are executed. A point mutation in the gene for protein that binds a light-sensitive pigment might cause the protein to interact with itself so that cluster of protein-bound pigments form in the membrane, making it sensitive for light. If light sensitivity helps the organism survive, maybe by avoiding predators or finding food more successfully, there would be positive selection pressure to keep this mutation. Additional mutations, maybe in genes that control where the pigment-binding molecules are expressed, or the shape of the structure holding these molecules, could lead to a structure that detects the direction of the light. These modifications would be retained if they proved helpful. Thus, it is possible that mutations accumulating over time could eventually produce an eye.

   Conversely, a specialized structure could develop suddenly if a large number of mutations occurred in the organism almost simultaneously, say as the result of mutagen exposure. Although statistically possible, the likelihood that a large number of random events happened almost simultaneously with such a favorable outcome seems very low.

Chapter 18

Biological Species Concept

Think about It

1. The ability to mate in nature; if two populations cannot interbreed, they represent different species.
2. (a) In captive populations, where breeding may occur between species that do not mate in nature; and (b) for bacteria and other asexual species, which do not (generally) have a sexual reproduction mode of replication.
3. B and C seem to be the same species based on the rRNA sequence analysis. Mating behavior is not informative for bacterial species.
4. top diagram—anagenesis, bottom diagram—cladogenesis
5. The two populations must cease interbreeding.

REPRODUCTIVE ISOLATING MECHANISMS

Think about It

1. some kind of geographical separation between the two populations
2. a. allopatric speciation

   b. an intrinsic isolating mechanism, such as species-specific behavioral mating rituals
3. See Table 18.1 in your textbook.
4. Sympatric speciation is the development of distinct species by intrinsic reproductive isolating mechanisms in the absence of geographic separation.

5. a. temporal isolation
   b. mechanical isolation
   c. hybrid infertility

THE CATEGORIZATION OF EARTH'S LIVING THINGS

Think about It

1. a standardized naming system for species
2. Your last name (surname) is your "genus-like" name, and your first name is your "species-like" name.
3. domain → kingdom → phylum → class → order → family → genus → species
   Tip: Some students find it easier to remember the order using a mnemonic device, such as Did King Philip come over from Germany Saturday?
4. morphology; geographical distributions of ancient and present-day organisms; DNA sequences
5. An analogy is a feature found in different organisms that has a similar function and appearance. Analogies are not based on the feature arising from a common ancestor; they arise by convergent evolution in unrelated species.
6. the order in which different species branched from a common ancestor
7. a. meat-eating
   b. whiskers
   c. Mimi and Georgia

Vocabulary Review

Place the following terms in the appropriate columns of the table. Some terms may be used in more than one column.

Speciation	Categorization
population	taxonomic system
allopatric	cladistics
sympatric	species
adaptive radiation	phylum
species	systematics
	Linnaeus

PRACTICE TEST

Compare and Contrast

cladogenesis/anagenesis

Both refer to modes of speciation (how species arise). However, in cladogenesis, new species branch off from a continuing species, whereas in anagenesis a new species arises without branching from the ancestral species.

extrinsic isolating mechanism/intrinsic isolating mechanism

Both refer to how different species may be reproductively isolated. However, extrinsic mechanisms are not part of the organisms' makeup (e.g., geographic separation by large rivers), whereas intrinsic mechanisms are part of the organisms' makeup (e.g., behavioral patterns that prevent mating with the wrong species).

phylum/kingdom

Both are levels of biological classification. However, a phylum is a level of classification below that of kingdom.

analogous/homologous

Both refer to features and the evolutionary relationships between organisms with these features. However, analogous features occur in unrelated groups, whereas homologous features are present in different species due to a shared evolutionary history.

Matching

a. (systematics)
b. (punctuated speciation)
c. (sympatric speciation)
d. (allopatric speciation)
e. (adaptive radiation)

Short Answer

1. Sympatric speciation occurs within the same geographic area or shared habitat, while allopatric speciation would take place in populations separated by some barrier.
2. Adaptive radiation is the appearance of a series of new species that arise due to the arrival of a single (ancestral) species in a new habitat.
3. The pace of evolutionary change is the major difference between the two models.
4. The number of shared derived characteristics is used to design phylogenetic trees.

Multiple Choice

1b, 2a, 3d, 4d, 5c, 6a, 7c, 8c, 9b, 10e, 11c, 12a

WHAT'S IT ALL ABOUT

1. **Type of Question**
   Variation on defend a position – asks you to apply your knowledge to describe an observation.
2. **Evidence Needed**
   What drives evolution? What advantages does complexity provide?
3. **Sample Answer**
   Evolution is driven by the force of natural selection on variations in organismal phenotypes. A phenotype that allows an organism to produce more offspring will become more common in successive generations as long as the environment remains stable. Complex structures allow for new or improved functions that allow organisms to succeed in different environments, such as ancient amphibians moving on to land. Complex structures can also allow organisms to adapt to changing environments; terrestrial animals can move to new places whereas fish are limited to their pond. Complexity is not better or worse than simple structures, but it does seem to allow organisms more options for dealing with the environment.

History of Life on Earth

Think about It

1. 4.6 billion years
2. decade, millennium, epoch, period, era
3. the impact of an asteroid on the Earth's surface
4. adaptive radiation of new species to fill emptied niches
5. E plants
   B E. coli
   A single-celled creatures in sulfur hot springs
   E fungi
   E roaches
6. an information molecule (e.g., DNA or RNA)
7. RNA can be both an informational molecule of heredity and a biological catalyst (ribozymes).
8. volcanoes, meteorites, deep-sea thermal vents
9. Compare and discuss possible answers with a classmate.

THE TREE OF LIFE

Think about It

1. Archaea and Eukarya (in spite of having different cell types)
2.

Domain	**Cell Type**
bacteria	prokaryotic
archaea	prokaryotic
eukarya	eukaryotic

3. Protista
4. Bacteria, approximately 3.5 million years ago
5. By photosynthesis; oxygen is used for cellular respiration by consumers and also provides the ozone layer for UV protection.

THE CAMBRIAN EXPLOSION

Think about It

1. The appearance of many new animal forms approximately 544 million years ago, and lasting approximately 5 million years. Almost all (29/30) modern phyla appeared then.
2. development of an ozone layer that provides protection from UV radiation
3. the plants
4.

Problem	**Solution**
water loss	waxy cuticle
reproduction	gamete production
gravity	vascular system

5. maturation of embryos within the parent
6. A fern; its vascular system allows the transport of water and nutrients against gravity.
7. photosynthesis, waxy cuticle, vascular system, seeds, flowers

8. to prevent water loss and protect from the Sun's rays
9. 1. plants
   2. insects
   3. amphibians
   4. reptiles
   5. mammals
10. An amniotic egg has a hard outer casing surrounding the developing embryo. It provides cushioning and an inner membrane system for nutrients and wastes. This egg allowed the move to land in the organisms that developed it.
11. by developing internal placental membranes, which serve the same nutrient and waste-exchange purpose

HUMAN EVOLUTION

Think about It

1. Africa
2. a different species (H. neanderthalensis vs. H. sapiens)

Vocabulary Review

Geological Time	Family Tree
epoch	bryophyte
era	amniotic egg
Permian	primates
Cambrian	Protista
continental drift	seedless vascular plant

PRACTICE TEST

Compare and Contrast

angiosperm/gymnosperm

Both are vascular plants that produce seeds. However, gymnosperms are nonflowering plants, whereas angiosperms are flowering plants.

Archaea/Bacteria

Both are kingdoms of unicellular organisms. However, Archaea include the extremophils.

epoch/era

Both are periods of time. However, eras are longer than epochs.

domain/phylum

Both are classifications of organisms. However, domains are larger (more inclusive) than phyla.

Matching

a. (lemur)
b. (accretion)
c. (Archaea)
d. (bryophyte)
e. (Bacteria)

Short Answer

1. in Africa, specifically in southern and eastern Africa
2. the development of a waxy cuticle and protection for the reproductive cells to prevent dehydration
3. Continental drift prevented the colonization of species from one geographic area to another. This in turn would have promoted the adaptive radiation of species into new niches, leading to new adaptations and new species.
4. The Permian Extinction caused the demise of approximately 96 percent of the existing species. This provided an opportunity for adaptive radiation by the survivors.

Multiple Choice

1d, 2e, 3c, 4e, 5c, 6b, 7e, 8c, 9a, 10d, 11c, 12d

BUT, WHAT'S IT ALL ABOUT?

1. **Type of Question**
   Variation on defend a position – asks you to apply your knowledge to describe an observation.
2. **Evidence Needed**
   This chapter – life is self-replicating
   Chapter 18 – For a living organism, need a self-replicating molecule that can also direct synthesis of other molecules (RNA).
   Chapter 5 — Also need a way to separate self from environment – membranes.
   Chapter 6 — Need a source of energy and the molecules to extract energy from the environment.
3. **Sample Answer**
   One of the first things needed for a living organism to evolve is a plasma membrane, to concentrate biologically active molecules and to separate them from the environment. I think this is the most time consuming process because it is limited by diffusion. We believe that the molecules important to living organisms were rare in the early oceans, so it probably took a very long time to capture enough of them inside of a membrane. Once inside a membrane, chemical reactions are more likely to happen. This means that a molecule capable of replicating itself, another criteria of a living organism, is more likely to do so if it is in an environment rich in its monomers. Because chemical reactions are more likely to happen, the evolution of pathways of reactions has also been speeded up, so a third feature of living

organisms can develop. Once these basic requirements of a living thing are met, self-replication, energy assimilation, interaction with the environment, then changes to the self-replicating mechanism can create new organisms.

Chapter 20

Viruses

Think about It

1. No; but a virus uses its own genetic machinery to take over a cell.
2. By infecting host cells and converting them to virus-making factories
3. The principle behind vaccination is to allow the body to develop an initial immune response against killed or attenuated viruses. Examples include vaccinations against measles and polio.

DOMAIN BACTERIA

Think about It

1. No, they are also responsible for nutrient cycling in the environment, they help protect our bodies from pathogens, and they synthesize vitamins in our intestines.
2. They convert nitrogen to a useful form, and they help in decomposition and sewage treatment.
3. (a) no nucleus; (b) no cytoskeleton, no membrane-bound organelles; (c) single chromosome; (d) asexual reproduction; (e) unicellular
4. No; most antibiotics are designed to be effective against bacteria, and viruses therefore are not affected by antibiotics.

DOMAIN ARCHAEA

Think about It

1. In conditions of (for example) high salt, high temperature, high pressure, and high or low pH. Venus is a lot hotter than Earth, so organisms there may resemble the archaeal theromophiles and have enzymes adapted to work at high temperatures.
2. Although their cells structurally resemble the Bacteria, Archaea seem to be more closely related to the Eukarya, and many of their genes resemble eukaryotic genes (although some also resemble bacterial genes).
3. Archaea are very similar (superficially) to Bacteria. Also, extremophiles are hard to grow (and study) under normal laboratory conditions.
4. Any place that has the correct conditions. Examples: cleaning car engines, ovens, chemical detoxification, acid runoff remediation, and so on.

KINGDOM PROTISTA

1. a. eukaryotic; b. sexual reproduction; c. moist environments; d. mostly microscopic; e. animal-like, plant-like, or fungi-like
2. Two billion years after Archaea and Bacteria
3. a. fungi-like; b. plant-like (algae)

4.

Life Stage	Activities	Cell Types
feeding	crawls along forest floor eating bacteria	single-celled
traveling	moves to a more fertile area of the forest	"slug" of 100,000 cells with front and back
reproduction	spore dispersal and development	stalk-like reproductive structure and spores

Vocabulary Review

Archaea	Bacteria	Protista
extremophile	prokaryotic	phytoplankton
prokaryotic	antibiotic	photoautotrophy
haploid	photoautotrophy	giardia
asexual reproduction	haploid	fungi-like
	asexual reproduction	eukaryotic
		diploid
		algae

PRACTICE TEST

Compare and Contrast

archaea/bacteria

Both are single-celled prokaryotes. However Archaea are more likely to be found in extreme environments, high temperatures or high salt content, for example, while bacteria are not.

autotrophs/heterotrophs

Both refer to modes of nutrition. However, autotrophs use the Sun's energy to synthesize organic molecules from inorganic molecules (like CO_2), whereas heterotrophs consume organic molecules.

phytoplankton/zooplankton

Both are types of protists. However, phytoplankton are autotrophs, whereas zooplankton eat phytoplankton.

viruses/protists

Both can be pathogenic, that is, disease-causing. However, viruses are acellular and therefore not considered to be living while protists are.

Matching

a. (algae)
b. (virus)
c. (protozoan)

d. (extremophiles)
e. (bacteria)

Short Answer

1. The Domain Bacteria is most similar in terms of organismal size and organization although archaea may also have some ties to the eukarya as well.
2. Photosynthetic aquatic organisms that belong to the Domain Prokarya or Kingdom Protista (algae) are known as phytoplankton.
3. Penicillin is an antibiotic, an anti-bacterial agent that is derived from cultures of the bread mold (fungus) *Penicillium chrysogenum.*
4. Pseudopodia, or "fake feet" are used by amoeboid organisms to move through the environment.

Multiple Choice

1d, 2e, 3d, 4d, 5d, 6e, 7a, 8e, 9d, 10b, 11c, 12a

BUT, WHAT'S IT ALL ABOUT

1. **Type of Question**
   Variation on defend a position – asks you to apply your knowledge to describe an observation.
2. **Evidence Needed**
   Why does any organism persist? As long as organisms successfully reproduce, that is adapt to changes in the environment and survive, they will persist unless something can drive their population to extinction, such as a natural disaster.
   Protists are generally single celled organism that can reproduce sexually or asexually. Like bacteria, they can exist in large populations, reproduce quickly and adapt rapidly to changes in the environment.
3. **Sample Answer**
   Organisms will persist as long as they can successfully reproduce and as long as no natural disaster drives the population to extinction. Natural selection doesn't specifically choose complex organisms over simple ones; it selects successful organisms. The protists, as a Kingdom, are remarkably diverse organisms. They can exist in large populations in aquatic or terrestrial environments. They can reproduce quickly. They have exploited many strategies for finding food. As a group, protists are adaptable organisms, so they persist.

Chapter 21

Think about It

1. protista, plantae, fungi, animalia
2. In humans, two nuclei join to form one. In fungi, the nuclei remain intact, forming a dikaryotic structure.
3. No, probably not. A fungal infection requires medication that is targeted to its life cycle.
4. See text.
5. Lichens are a composite of fungus and algae or bacteria.

Think about It

1. **bryophyte**
   moss growing next to a driveway

   seedless vascular plant
   fern growing in a window

   gymnosperm
   pine tree in a park

   angiosperm
   crocuses blooming in a lawn in early spring.
2. A seed is a developing embryo, and its food supply is enclosed in a durable case. The seed-producing plants are successful because the seeds can be spread over long distances.
3. Plants have flowers to attract pollinators and thereby increase reproductive success.
4. The endosperm provides nourishment to the plant embryo. Edible endosperm includes rice, coconut, and wheat.
5. Fruit aids in seed dispersal by inducing animals to eat the fruit and associated seeds. Cherries and cucumbers are both fruit, although cucumber is often thought of as a vegetable.
6. a. multicellularity; b. vascular system; c. seeds; d. flowers

Vocabulary Review

Plants	Fungi
Eukaryotic	hypha
Chloroplast	Eukaryotic
Spores	Rhizoids
Seeds	Yeast
Endosperm	Molds
Diploid	haploid
vascular system	

PRACTICE TEST

Compare/Contrast

autotroph/heterotroph
Both refer to modes of nutrition. However, autotrophs use the Sun's energy to synthesize organic molecules from inorganic molecules (like CO2) whereas heterotrophs consume organic molecules.

cup fungi/imperfect fungi
Both are categories of fungi. However the term "imperfect fungi" (or mitosporic) is used for species that have not (yet) been observed to reproduce sexually.

sporophyte/gametophyte
Both terms refer to the alternating generations of plants. However, the gametophyte produces haploid structures while those of the sporophyte are diploid.

angiosperm/gymnosperm

Both are seed-producing plants. However, angiosperms have flowers and gymnosperms do not.

Matching

a. (gymnosperms)
b. (mycorrhizae)
c. (dikaryotic)
d. (ferns)
e. (chloroplasts)

Short Answer

1. Bryophyte sperm must swim through water in order to fertilize the egg.
2. A lichen is a symbiotic association of fungi with either algae or bacteria.
3. Gymnosperms have both seeds and a vascular system.
4. Plants use nectar as an attractant for animal pollinators.

Multiple Choice

1d, 2d, 3e, 4c, 5a, 6c, 7e, 8a, 9a, 10e, 11a, 12a, 13a

BUT, WHAT'S IT ALL ABOUT

1. **Type of Question**

   Variation on defend a position – asks you to apply your knowledge to describe an observation.

2. **Evidence Needed**

   What is the relationship between animals and angiosperms? What advantages to animals provide to angiosperms and vice versa? What disadvantages exist for both organisms? What drives the evolution of organisms?

3. **Sample Answer**

   The relationship between animals and angiosperms could only have evolved if it provided a selective advantage for survival for both organisms. Producing flowers and fruit requires a lot of energy, but the disadvantage of needing to assimilate more energy is offset by the advantage that animals are more precise pollinators and more effective at spreading seeds than the wind or water. Animals that adapt to eating nectar or fruit gain the advantage of a high-energy food source that doesn't run away, which usually outweighs the disadvantage of relying on a food source than can change in quality of quantity because of the climate. Because many species of angiosperms and animals exist, we can conclude that the relationship between these two groups is evolutionarily advantageous.

Chapter 22

Defining an Animal

Think about It

1. Sponges are multicellular heterotrophs lacking cell walls that go through a blastula stage during development.

2. No, the difference is developmental; protostomes have a ventral nerve cord, whereas deuterostomes have a dorsal one.
3. Porifera (1,10); Mollusca, Annelida, Arthropoda (5,6,7); Echinodermata, Chordata (8,9); Cnidaria, Platyhelminthes, Nematoda (2,3,4)

NONCOELOMATES: PHYLA PORIFERA AND CNIDARIA

Think about it

1. Sponges have microscopic pores between the cells to allow water to wash in nutrients and wash out wastes.
2. Both go through a polyp stage during development.
3. tentacles for stinging prey (cnidarians) v. filtering water for food (sponges)

PROTOSTOMES

Think about it

1. develops from mesoderm tissue in embryo; tissue lacking in Cnidarians and Poriferans
2. Flatworms need to have all their cells within reach of air and nutrients.
3. As they grow, arthropods must molt; a new shell grows beneath the old one, which must be discarded when it gets too small.
4. Yes; arthropods have segmented bodies and legs, and vertebrates have a segmented backbone.
5. Insects coevolved with plants to adapt to feeding on angiosperms.
6. to have three developmental layers—endoderm, mesoderm, and ectoderm
7. a. S,
   b. R,
   c. F,
   d. R,
   e. F
8.

Cephalopod	Gastropod	Bivalve
squid octopus nautilus	snails slugs	oysters mussels clams

DEUTEROSTOMES

Think about It

1. presence of a vertebral column; flexible column of bones extending anterior to posterior
2. improved ability to capture and consume prey
3. The amniotic egg protects embryo from drying out, decreases its dependence on wet environment for survival.
4. (a) mammary glands; (b) constant internal temperature; (c) hair; (d) internal development of eggs

5. Mammals need to consume more food to maintain a high enough metabolic rate.
6. Cephalochordata and Urochordata lack backbones.

Vocabulary Review

Chordata	Nonchordates
coelom	coelom
bilateral symmetry	Mollusca
tissues	bilateral symmetry
mammary glands	tissues
monotreme	Arthropoda
vertebral column	radial symmetry
blastopore	blastopore
hair	invertebrate
	diploblastic
	Echinodermata
	nematodes

Compare and Contrast

protostome/deuterostome

Both types of animals possess a coelem. However, the orientation of the developing embryo differs such that protostomes have a ventral nerve cord and dorsal heart whereas deuterostomes have a dorsal nerve cord and ventral heart.

roundworms/flatworms

Both animals have bilateral symmetry and organs. However, flatworms lack a central body cavity whereas roundworms have a "pseudo-coel," a primitive body cavity.

ectothermy/endothermy

Both are ways of regulating internal temperature. However, in ectothermy the ambient temperature determines an animal's internal temperature whereas in endothermy the internal temperature is regulated by the animal's metabolism.

triploblastic/diploblastic

Both terms describe the blastula. However, a triploblastic blastula has three germ layers; the endoderm, mesoderm and ectoderm whereas a diploblastic blastula has only two germ layers, the endoderm and ectoderm.

Matching

a. (reptiles)
b. (porifera)
c. (Mollusca)

d. (marsupials)
e. (Platyhelminthes)

Short Answer

1. Multicellular, heterotrophic, no cell walls, blastula. Only animals have a developmental stage known as the blastula.
2. Phylum is a term used to group organisms by their shared characteristics. It is broader than a family but narrower than a kingdom.
3. the presence of the notochord, a nerve cord (dorsal), postanal tail, and pharyngeal slits
4. Millipedes, centipedes, and insects are all members of this group.

Multiple Choice

1a, 2c, 3e, 4e, 5a, 6c, 7b, 8b, 9d, 10e, 11e, 12a

BUT, WHAT'S IT ALL ABOUT

1. **Type of Question**
   Variation on defend a position – asks you to apply your knowledge to describe an observation.
2. **Evidence Needed**
   What is the hierarchy of life (hint, begins with atoms…)? What does this hierarchy describe? What does phylogenetics describe?
3. **Sample Answer**
   Phylogenetics reflects a hierarchy of functions that result developing a hierarchy of structures. Classification in Kingdom Animalia reflects gains in structures that support new functions; for example, members of Phylum Porifera are little more than tubes of cells (tissues) that can absorb nutrients from water. By gaining body projections (organs — tentacles), members of Phylum Cnidaria gain the function of prey capture. Subsequent additions of bilateral symmetry, a body cavity, and so forth each allow new functions for the organism. Just as the structure of a molecule directs its function, so does the structures found within organisms direct their expanding array of functions.

Chapter 23

The Structure of Flowering Plants

Think about it

1. Gymnosperms (3); angiosperms(4); bryophytes(1); seedless vascular(2)
2. Please refer to your text to label the plant parts.
3. Please refer to the text.
4. R – anchoring; S – photosynthesis; S – reproduction; R – water absorption
5. The roots need to provide sufficient surface area to absorb enough water for the plant's needs.

6.

	Water Transport	Food Transport
System in Plant	xylem	phloem
Direction of Transport	roots to leaves	throughout the plant
Cells Involved	tracheids, vessel elements	companion cells, sieve cells

THE FUNCTION OF FLOWERING PLANTS

Think about It

1. M – stamen F – style F – stigma M – anther
   F – carpel M – pollen M – filament
2. The mark will be at the same height from the ground, because the new tree height will have grown from the apical meristem at the treetop.
3. Phototropism is the growth of a plant toward a light source.
4. a. Gravitropism. Currently, it is unclear how plants sense and respond to gravity, because the original hypothesis postulating the use of amyloplasts does not appear to hold for those plants that can sense gravity but lack amyloplasts.
   b. thigmotropism
5. Ragweed is a long-night plant that responds to a photoperiod of shorter days and longer nights. Lettuce is a short-night plant, responding to a photoperiod of longer days and shorter nights.

Reproduction	Transport	Signals
Flowers	roots	Photoperiodism
seeds	phloem	Thigmotropism
pollen	xylem	Phototropism hormone
megaspore	vascular system	

PRACTICE TEST

Compare and Contrast

angiosperm/gymnosperm

Both are seed-producing plants. However, angiosperms have flowers, while gymnosperms do not.

stamen/carpel

Both are reproductive structures found in flowers. However, the stamen is the male reproductive structure, while the carpel is the female reproductive structure.

root/stem

Both are major systems of plants. However, the roots are below ground, while the shoots are found above ground.

xylem/phloem

Both are part of a plant's vascular system. However, xylem transports water, while phloem transports food.

Matching

a. (cuticle)
b. (stomata)
c. (thigmotropism)
d. (transpiration)
e. (bryophyte)

Short Answer

1. They provide passage for the phloem sap.
2. Plants provide not only habitat for the snake to live, hide, and hunt in but also food for her prey.
3. Plants use structural and chemical defenses like thorns and toxins.
4. Deciduous describes plants that demonstrate a coordinated seasonal loss of leaves.

Multiple Choice

1b, 2d, 3e, 4d, 5b, 6a, 7a, 8c, 9c, 10b, 11c, 12c

BUT, WHAT'S IT ALL ABOUT

1. **Type of Question**
   A two-part question, first you need to show you know what evolutionary changes differentiate seedless plants from conifers and angiosperms and then, by analogy, indicated what evolutionary changes have defined success for the angiosperms.
2. **Evidence Needed**
   From Chapter 21 – The ancestors of the gymnosperms and angiosperms are seed bearing, so the first major evolutionary change was the development of pollen and seeds. Remember that in the simpler plants we have a gametophyte generation that produces undifferentiated gametes. In the seed bearing plant, these gametes have become specialized into pollen and megaspore; the gametes have become more specialized in function and more adaptable to a non-aqueous environment. The second evolutionary development was the production of fruit to surround the seeds, so there had to be a change in the developmental instructions to the cells around the seeds to increase food storage. A point raised in this chapter also helps us understand the success of the angiosperms, that is the changes in the plant's phenotype to attract pollinators.
   From Chapter 17 – The 5 agents of microevolution include mutation, gene flow, genetic drift, nonrandom mating, and natural selection. The changes from seedless to seed bearing probably depended mostly upon mutation, although it is possible that primitive plants could have acquired new genes as a result of polyploidy. The forces of natural selection would have tested these mutations.
3. **Sample Answer**
   You should see the pattern now – define the changes and describe how these changes could have come to be.

Categorizing Flowering Plants

Think about It

1. One way to classify plants is based on the length of the life cycle; another is based on the number of embryonic leaves. A home gardener would be more likely to use the length of the life cycle (annual, perennial, etc.).
2. Monocots and dicots differ in the number of embryonic leaves, root system, leaf structure, and arrangement of vascular bundles.

PLANT CELLS AND TISSUE TYPES

Think about It

1. S (contains cellulose); P (most abundant cell type); S (thick-walled cells); C (midway between the other two types)
2. Sclerenchyma cells are dead at maturity.
3. Parenchyma cells are the basic starting cells from which other cells differentiate.
4.

Tissue Type	Function(s)
Dermal	Makes the outside covering of the plant
Vascular tissue	Enables transport of water and food
Meristem	Makes up material responsible for growth
Ground tissue	Most basic plant functions ("everything else")

5. P/S (trees); P (herbs); P (orchids)
6. Apical meristems are found at the root and shoot tips. They continue to divide to produce new plant cells.
7. Yes, it will continue to grow because a lateral bud will become activated.
8. Please refer to the text.
9. Vascular cambium, which produces secondary xylem and phloem, and cork cambium, which produces the outer layer of the plant.
10. secondary xylem
11. This removes the living phloem, so food transport is prevented.
12. tracheids and vessel elements
13. Water is pulled by the force of water evaporation through evaporation from the leaves.
14. Phloem sap contains mostly sucrose and water. Transport is from a source (site of production or storage) to a sink (site of usage).
15. arteries and veins

SEXUAL REPRODUCTION IN FLOWERING PLANTS

Think about It

1.

Generation	Ploidy
Sporophyte	diploid
Gametophyte	haploid

2. There are two sperm, and each fertilizes different cells. One fertilization produces the embryo and the other produces the endosperm. Please refer to the text for an illustration of double fertilization.
3. seed (embryo with food source in a protected case); fruit (tissue surrounding the seeds which develops from the ovary); pollen (contains male gametes)
4. germination of the seed
5. The water source is produced first because the root structure emerges from the seed first.

Vocabulary Review

Growth	Transport	Reproduction
Meristem	xylem	sporophyte
Phelloderm	phloem	endosperm
cork cambium	tracheid	ovule pollen
cork	sieve elements	gametophyte

PRACTICE TEST

Compare and Contrast

pollen/endosperm

Both are associated with angiosperm reproduction. However, pollen contains the male gametes, and the triploid endosperm is contained within the seed and serves as a nutrient source for the embryo.

monocot/dicot

Both are types of angiosperms. However, monocots have one embryonic leaf, while dicots have two embryonic leaves.

primary/secondary growth

Both are types of plant growth. However, secondary growth occurs in woody plants and involves an increase in girth, while primary growth occurs in all plants and is responsible for increases in height of the plant.

cork cambium/vascular cambium

Both are meristems involved in secondary growth. However, cork cambium produces the outer layer of woody plants, while the vascular cambium produces secondary xylem and phloem.

Matching

a. (seed coat)
b. (companion cells)
c. (fruit)
d. (pith)

Short Answer

1. Ovaries become fruit, so the statement is really saying "An apple a day keeps the doctor away."
2. The meristem is involved in growth.
3. Xylem flows in through the roots and out (via transpiration) through the leaves.
4. the diploid generation of a plant

Multiple Choice

1e, 2b, 3c, 4c, 5c, 6e, 7c, 8c, 9c, 10c, 11c, 12d

BUT, WHAT'S IT ALL ABOUT

1. **Type of Question**
   Variation on defend a position – asks you to apply your knowledge to explain an observation.
2. **Evidence Needed**
   Water Transport
   Water is pulled into the root to replace water lost by the leaves during transpiration. If more water surrounds the root than can be removed by transpiration, an aqueous environment that favors bacterial growth surrounds the cells.
   Food Transport
   A high water concentration in the root cell would dilute the sugar flowing in from the phloem and decrease the osmotic pressure so less water flows back into the xylem. Water will only move up the xylem if the rate of transpiration is greater than the rate of absorption, so slower osmosis of water out of the roots would slow sap flow. All of this might affect the energy balance of the plant, making it more vulnerable to infection.
3. **Sample Answer**
   You should see the pattern now – define the changes and describe how these changes could have come to be.

Chapter 25

General Characteristics of Humans

Think about It

1. a. – A
   b. – P
   c. – P
   d. – A
2. a. ventral
   b. dorsal
3. a. A cooking pot is ectothermic; the stove is endothermic.

b. Reptiles are ectothermic. This means that they rely on an external source of heat to keep their bodies warm.

4. molecules → cells → tissues → organs → organ systems
5. Cells are the smallest unit of life. The heart is an organ, and all the blood vessels in the body make up an organ system.

TISSUES, ORGANS, AND ORGAN SYSTEMS

1.

Organ System	Role
Lymphatic	Collect and deliver fluid and check it for infection
Cardiovascular	Transport of nutrients and waste products
Urinary	Get rid of waste products
Integumentary	Protection against desiccation and invasion by foreign organisms (like bacteria)

2. a. Blood is connective tissue, and the linings of blood vessels are epithelial.
   b. nervous
   c. muscle
   d. epithelial
   e. connective (adipose)
   f. connective
3.

Tissue	Function(s)	Specific Example
Epithelial	Covers internal and external surfaces	Epidermis of the skin
Connective	Provides support and protection	Bone
Muscle	Produces contraction	Heart muscle
Nervous	Enables rapid communication	Brain

4. Waterproofing to prevent water loss.
5. ST – heart muscle and biceps of the arm; SM – muscles around the bladder
6. Please refer to the text for an illustration of a neuron. The structures you draw should include the nucleus, the cell body, the axon, and the dendrites.

THE INTEGUMENTARY SYSTEM

1. Sebaceous glands secrete sebum (lubrication). Sweat glands secrete sweat, which acts to cool the body's surface.
2. the arrector pili muscles
3. apocrine
4. a. Bone marrow is loose connective tissue.
   b. Blood is fluid connective tissue.
   c. Skin is part of the integumentary system (epithelial tissue).

5. a. osteocytes (mature cells), osteoblasts (immature cells) and osteoclasts (recyclers)
   b. 8 months osteoblasts
   28 years osteocytes
   48 and 68 years osteocytes and osteoclasts
   c. osteoclasts, because they recycle bone to release calcium
6. Because joints are structures that balance the need for strength with the need for flexibility, they are thus inherently weaker than bones.
7. to provide as much cushioning as possible in a compact space, and the ability to change the shape of the cushioning as the joint moves
8. a. It shortens.
   b. By shortening while attached at both ends to objects, those can move relative to one another (be brought closer together, for example).
   c. by forming a meshwork of sarcomeres, for example around the heart
9. a. 1. Myosin heads attach to actin filaments.
   2. Pull actin filament to middle of sarcomere.
   3. Myosin heads detach from actin filament.
   b. ATP is the energy source for muscle contraction.

Vocabulary Review

Body Support and Movement	Coordination, Regulation	Transport and Exchange with the Environment
muscle	nervous	lymphatic
skeletal	respiratory	urinary
nervous	endocrine	integument
	digestive	cardiovascular

PRACTICE TEST

Compare and Contrast

ectotherm/endotherm

Both refer to temperature regulation mechanisms. However, ectotherms rely on an external heat source to heat their bodies, while endotherms generate heat from within their bodies.

exocrine/endocrine glands

Both are glands that secrete substances. However, exocrine glands have ducts, while endocrine glands do not.

basement membrane/ground substance

Both describe components of tissue. However, the basement membrane is a network of protein fibers to which epithelial cells attach, while the ground substance is the fluid-like material in which the fibers of connective tissue are found.

organelle/organ

Both are structures in the body. However, an organelle is a structure found within cells that carries out a specific function, while an organ is a collection of cells and tissue that work together to carry out a specific function.

Matching

a. (heart)
b. (skin)
c. (bone)
d. (nervous)
e. (epithelial)

Short Answer

1. Connective tissues are categorized by their extracellular material or matrix.
2. Cardiac muscle is found in the heart, skeletal (or striated) muscle moves the body through space, and smooth muscle moves material within the body.
3. Exocrine glands release materials through ducts. Examples are the sweat and sebaceous glands of the skin.
4. Red bone marrow generates blood and immune cells; yellow bone marrow stores fat.

Multiple Choice

1d, 2d, 3c, 4a, 5d, 6d, 7b, 8c, 9d, 10a, 11c,12d

BUT, WHAT'S IT ALL ABOUT?

1. **Type of Question**
   Two compare-contrast questions which are connected because the goal is to determine if similar types of tissues have similar types of functions in plants and animals.
2. **Evidence Needed**
   Make a table! Unique functions in each organism are in *italics*

Plants	Animals
Dermal Tissue (epidermis) Functions: protective, gas exchange, *water absorption*	Epithelial Function—protective, exchange w/ environment, *exocrine & endocrine glands secreting hormones*
Ground Tissue Function—bulk of plant—structure, storage, *photosynthesis*	Fibrous Connective Tissue, Muscle Tissue, Loose Connective Tissue, Adipose Functions—bulk of animal—structure, storage, *packing material*
Vascular Tissue Xylem & phloem—water & food flow	Epithelial tissue makes the vascular vessels Fluid Connective Tissue transports water & food *Nervous tissue*

The Nervous System

Think about It

1. Please refer to the text.
2.

Interpreting these words as you read them	CNS
Carrying the "pain" message from your burned finger to your brain	Afferent nerves
Carrying messages to and from the extremities	PNS
The nerves that cause you to pull your hand away from a hot stove	Efferent nerves

3. to surround and support neurons and provide protection
4. Afferent division senses stimuli; efferent responds to stimuli.
5.

Neuron Type	Role
Sensory	Sense internal and external conditions
Motor neurons	Carry information from the CNS to tissues
Interneurons	Interconnect neurons within the brain and spinal cord

6. a. Myelin speeds up nervous transmission.
   b. Both are composed of neuroglia, the supporting cells of the nervous system.
7. motor neurons — turn head
   sensory neurons — smell rose
   interneurons and motor neurons — pick up rose
8. The charge distribution is called the membrane potential. Please refer to the text for an illustration. The neuron is relatively negative inside. The concentration of $Na+$ is high outside, while the concentration of $K+$ is high inside.
9. $Na+$ ions flow in when the $Na+$ gates open, then $K+$ flows out when the $K+$ gates open. The membrane potential changes in one section of the membrane, then the ion gates are triggered to open in the next section of the axon membrane.

10.

Component	Role in Transmission between Neurons
Presynaptic neuron	Releases neurotransmitter into the space between neurons
Synaptic cleft	Space between neurons across which a neurotransmitter carries the message
Postsynaptic neuron	Receptors bind the neurotransmitter and the cell responds by opening the Na+ gates and initiating an action potential

11. 1. presynaptic neuron
    2. neurotransmitter
    3. synaptic cleft
    4. postsynaptic neuron
12. Neurotransmitters communicate between neurons at synapses. They do not all send the same message, for example acetylcholine is stimulatory while dopamine is inhibitory.
13. a. action potential
    b. synaptic transmission
14. due to reflex arc running from the sensory neuron in your hand through the spinal cord to a muscle, which contracts in response
15. The parasympathetic division is active as you relax with your book, and the sympathetic branch is active as you combat a frightening situation.
16. Most organs receive instructions from both systems to help maintain balance and stability.
17. studying cerebrum
    breathing medulla oblongata
    listening midbrain
    watching midbrain
18. because the medulla oblongata controls the basic functions critical to life, such as breathing, heart activity, and digestion
19. 1. PNS
    2. spinal cord
    3. brain
    4. somatic
    5. afferent
    6. sympathetic

THE ENDOCRINE SYSTEM

1. endocrine glands, which have no ducts
2. Steroid hormones. A certain amount of cholesterol is necessary to make these essential hormones (which include the sex hormones).
3. They travel through the circulation.
4. Please refer to the text for the action of amino-acid-based hormones. Steroid hormones can pass though the membrane (unlike the amino-acid hormones).

5. Hormones alter cellular activities by changing the production or shape of enzymes or structural proteins.
6. Positive feedback involves an increase in the response as a result of the stimulus. The response drives the system away from a level or steady state. Negative feedback results in a decreased response as a result of the stimulus. This serves to maintain a level or balance in the system.
7. Because negative feedback removes the stimulus instead of increasing it. This is important if the stimulus is something like low blood sugar.
8. homeostasis
9.

Pituitary	Synthesis of own hormones?	Hormone release controlled by hypothalamus?
Posterior	No	Yes
Anterior	Yes	Yes

10. LH
11. anterior pituitary

Vocabulary Review

Nervous System	Endocrine System
axon	Gland
homeostasis	prolactin
neuroglia	pituitary
neurotransmitter	steroid
pituitary	negative feedback
pons	
hypothalamus	
negative feedback	
synapse	

Compare and Contrast

nerve/neuron

Both are integral components of the nervous system. However, a nerve is a bundle of axons, while a neuron is a single nervous-system cell.

steroid/amino-acid hormones

Both are involved in endocrine signaling and cause responses in target cells. However, amino-acid hormones bind to receptors on cell surfaces, while steroid hormones enter cells to exert their effects.

pituitary/hypothalamus

Both are nervous system structures involved in endocrine control. However, the pituitary controls the release.

axon/dendrite

Both are part of a neuron. However, the dendrite receives information, and the axon transmits information from the cell body to the synaptic terminal.

Matching

a. (Autonomic Nervous System)
b. (hormone)
c. (action potential)
d. (synapse)
e. (pituitary)

Short Answer

1. The difference is between cell-to-cell contact versus using the circulatory system to carry messages.
2. Rods and cones are photoreceptor cells found in the eye. Rods respond primarily to low light situations while cones respond to bright light and therefore, color.
3. Homeostasis is maintained primarily by negative feedback.
4. If the eyeball is too long, the image focuses in front of the retina instead of on it. This is called nearsightedness. In an eyeball that is too short the image focuses behind the retina, causing farsightedness.

Multiple Choice

1b, 2c, 3e, 4d, 5b, 6d, 7d, 8c, 9d, 10d, 11c, 12d, 13c, 14d, 15d, 16d, 17d, 18a

BUT, WHAT'S IT ALL ABOUT?

1. **Type of Question**
   Analysis type question – think about how we see and try to figure out how being blind for a long time affects the system.
2. **Evidence Needed**
   1. All sensory neurons deliver information to the brain
   2. Visual system: detect light, focus image
      convert image into neural signal
      brain integrates neural signal
3. What does it mean to integrate a signal? Combine it with information from other senses and experience. Experience is knowing that an item shaded on the bottom is raised above the surface—if you've never learned that, will it make sense the first time you see it?

The Immune System

1. The specific immune system fights a particular invader, whereas the nonspecific system is a form of generalized protection against any invader. They are similar in that both defend the body against invading foreign substances.
2.

Physical barriers	Cellular	Molecular	Physiological responses
epithelium	phagocytes	interferon	fever
keratin layer on skin	NK cells	complement	inflammation
secretions			
hair			
mucus			

3. due to the keratin layer, sebaceous secretions, and hair
4. a cell that ingests and digests foreign particles and cellular debris
5. natural killer cells
6. P – infant drinks mother's milk
   A – polio vaccination
   P – injection of antibodies
7.

	Cell Types	Antibody Production?	Direct Contact with Target Cell?
Antibody-mediated	B cells	Yes	No
Cell-mediated	T cells	No	Yes

8. a. For chicken pox, memory cells persist after the first infection. These memory cells are rapidly activated upon subsequent infections.
   b. You get many colds, because each cold virus is slightly different, so that memory cells from one cold do not protect against the next cold virus.
9. a. Plasma cells produce antibodies.
   b. Memory cells are long lasting in the body.
10. helper T cells and cytotoxic T cells
11. because they produce an inappropriate immune response
12. opportunistic infections
13. It is a measure of the number of helper T cells.
14. the HIV virus
15. They block a viral enzyme that is critical for HIV replication within infected cells.
16. Use your text to complete the flowchart.

PRACTICE TEST

Vocabulary Review

Specific			Non-specific		
T cell	antibody	antigen	histamine	sweat glands	antigen
vaccine	helper T cell immunity		first line	macrophage	inflammation
B cell	macrophage				second line

Compare and Contrast

specific/nonspescific defenses

Both are parts of the body's defenses. However, the specific system is protection against particular invaders, while the nonspecific system is generalized for any invader.

B cell/T cell

Both are white blood cells involved in immune system function. However, B cells differentiate to produce antibodies, while T cells are involved in cell-mediated immunity. specific defense/nonspecific defense

histamine/cytokine

Both histamine and cytokine are molecules involved in the immune response. However histamine generates the inflammatory response while cytokines also play a role in stimulating the specific immune response.

dendritic cells/macrophages

Both dendritic cells and macrophages play important roles in the early stages of the immune response. However macrophages alert the body infection as part of the nonspecific defenses while dendritic cells are part of the specific responses.

plasma cells/B cells

Both plasma cells and B cells are involved in the antibody-mediated immune response. However, while B cells give rise to plasma cells, it the plasma cells that actually produce circulating antibodies.

Matching

a. (interferon)
b. (allergen)
c. (antigen)
d. (dendritic cells)
e. (B cell)
f. (Helper T cell)

Short Answer

1. Autoimmune disorders are caused by the immune system attacking cells within the body, like the joints (rheumatoid arthritis) or the nervous system glial cells (multiple sclerosis).

2. A vaccine is a substance that generates an "immune-like" response to a specific type of molecule. The body is then well-primed for exposure to the real antigen and able to defend itself against attack.
3. Inflammation is a non-specific response to invation and includes the symptoms of heat, redness, pain and swelling.
4. While they are both caused viruses, their treatments differ. Polio can be prevented by vaccination with attenuated or killed viral particles while the nature of HIV is such that it makes it much more difficult if not impossible to design and effective vaccine.

Multiple Choice

1b, 2c, 3d, 4d, 5d, 6d, 7a, 8e, 9b, 10d, 11e, 12b

BUT, WHAT'S IT ALL ABOUT?

1. **Type of Question**
   Defend a position, and the question could go either way.
2. **Evidence Needed**
   a. Specific immune system depends upon exposure to a non-self molecule (antigen)
   b. Antigen binding to B cells elicits antibodies, to macrophages activates T cells
   c. In both cases, the B or T cell has to be able to recognize the antigen, meaning have the receptor

   So

   Early exposure to antigens shouldn't matter because if you have circulating B cells that recognize a self-protein, you can trigger an autoimmune reaction no matter how many other antigens your body sees.

   But maybe,

   Early antigen exposure does protect you because exposure to a large number of antigens creates large numbers of circulating memory cells so that the percentage of B cells sensitive to self-antigens are diluted in the B cell population, making it very unlikely that a B-cell carrying a self-protein receptor will ever locate its antigen

Chapter 28

The Cardiovascular System and Bodily Transport

Think about It

1. Plasma – fibrinogen, albumin, globulins
   Cells – RBC, WBC, platelets
2. Plasma is blood without the formed elements. Serum is the fluid that remains after clotting of the plasma.
3. Because large animals need a mechanism to ensure that oxygen transported by the blood can reach all the cells of the body.
4. Please refer to the text.
5. capillary — single-cell epithelium lining
   artery — three layers
   vein — three layers with valves

6. Two chambers receive blood into the heart and two pump blood out of the heart. Two handle oxygenated blood and two handle unoxygenated blood.
   (i) Right atrium receives unoxygenated blood from the body.
   (ii) Right ventricle pumps unoxygenated blood to the lungs.
   (iii) Left atrium receives oxygenated blood from the lungs.
   (iv) Left ventricle pumps oxygenated blood to the body.
7. The systemic circulation drops off oxygenated blood at the cells of the body, while the pulmonary circulation picks up oxygen from the lungs.
8. The left ventricle is strongest because it has to pump blood to the entire body (the greatest distance).
9. Coronary arteries – they normally supply blood and oxygen to the heart muscle.
10. heart → arteries → arteriole → capillaries → venules → veins
11. The exchange takes place at capillary beds. The skeletal musculature helps move blood through the veins.
12. Osmosis permits water to flow back into the capillary beds at the venule end (which was forced out at the arteriole end by the higher blood pressure). In this way, most of the water is returned to the blood.
13.

Arterioles	Venules
O_2 glucose red blood cells	CO_2 wastes nitrogenous wastes red blood cells

14. Please refer to the text.
    1. right atrium
    2. right ventricle
    3. pulmonary arteries
    4. pulmonary veins
    5. left atrium
    6. left ventricle
    7. aorta

THE RESPIRATORY SYSTEM

Think about It

1. Humans inhale air due to the expansion of the chest cavity by the rib cage and muscular contractions of the diaphragm.
2. alveoli; hemoglobin
3. Blood needs to carry carbon dioxide in order to remove carbon dioxide (a product of cellular respiration) from the tissues. Seventy percent is carried as bicarbonate ions; 23% is bound to hemoglobin.
4. to provide sufficient surface area to exchange enough oxygen for the needs of every cell in the body

5. lung — high oxygen, low carbon dioxide
   pulmonary capillaries — high carbon dioxide, low oxygen
6. Please refer to the text.
   1. larynx
   2. esophagus
   3. trachea
   4. bronchioles
   5. lung
   6. diaphragm

THE DIGESTIVE SYSTEM

Think about It

1. ingestion → digestion → absorption → excretion
2. Ingestion — getting food into the digestive system
   Digestion — breaking the food down into monomers
   Absorption — taking up the monomers to use as cellular building blocks
   Excretion — elimination of any remainder after digestion and absorption
3. carbohydrate (chemical) digestion, as well as general mechanical digestion
4. A cow would have prominent bicuspids and molars because it has a vegetarian diet. A wolf would have prominent incisors and cuspids (because it is a carnivore).
5. the pharynx and the stomach
6. Pepsin digests proteins; hydrochloric acid maintains the low pH, kills microorganisms, and helps break down food structures.
7. stomach → duodenum → jejunum → ileum → large intestine
8. the completion of digestion and absorption of water and nutrients
9. bile, which is produced by the liver and aids in the digestion of fat
10. The bacteria assist the colon in completing the digestion of proteins and carbohydrates.
11. It is reabsorbed by the intestine.
12. Vitamins A and D are fat soluble, so can be stored in the body and accumulate to potentially dangerously high levels. Vitamin C is water soluble, so any excess can be excreted in the urine.
13. The protein is digested in the stomach by pepsin and in the intestine by pancreatic secretions. The fat is emulsified by bile in the small intestine and broken down by pancreatic lipase. Carbohydrates are digested in the mouth by salivary amylases and in the intestine by amylases. Vitamins are absorbed in the small intestine.

THE URINARY SYSTEM

Think about It

1. The kidney produces urine, the ureter carries urine to the bladder for storage, and the urethra eliminates the urine from the bladder (and thus from the body).
2. All are functions of the urinary system.

3. A filtrate derived from the blood enters the Bowman's capsule from the glomerulus (branches from the renal arteriole). Urine leaves each nephron via the collecting duct to eventually leave the body through the urethra.
4. to retain valuable molecules, such as amino acids, vitamins and glucose
5. Once the filtrate reaches the collecting duct, it is referred to as urine.
6. Please refer to the text.
   1. Bowman's capsule
   2. glomerulus
   3. proximal tubule
   4. distal tubule
   5. vasa recta
   6. collecting duct/tubule
   7. Loop of Henle (ascending)
7. Kidney stones are deposits of minerals within the collecting ducts or ureters. They cause painful blockages of urine flow.
8. The female urethra is shorter than the male urethra, so bacteria can more easily travel to the female bladder.
9. The internal sphincter exerts involuntary control over discharge of urine from the bladder. The external sphincter is found in the urethra and is responsible for voluntary control.
10. The bladder increases in volume, forcing the internal sphincter to open and leading to relaxation of the external sphincter.

Vocabulary Review

Cardiovascular	Pulmonary	Digestive	Urinary
erythrocytes	intercostal	pancreas	kidneys
platelets	muscles	energy	blood volume
gas exchange	gas exchange	absorption	regulation
venules	alveoli	duodenum	urethra
			pH balance

PRACTICE TEST

Compare and Contrast

red blood cells/white blood cells

Both are part of the formed elements of blood. However, RBCs carry oxygen using hemoglobin, while WBCs fight foreign invaders.

artery/vein

Both carry blood in the circulatory system. However, arteries carry blood away from the heart, while veins carry blood toward the heart and have valves.

mechanical digestion/chemical digestion

Both are critical to break down food. However, mechanical digestion is tearing of the food, while chemical digestion involves breaking the covalent bonds within food molecules.

alveoli/arterioles

Both are involved in gas exchange. However, alveoli are sacs within the lungs, and arterioles are small blood vessels that carry blood to the capillary beds for gas exchange.

ureter/urethra

Both are part of the urinary system. However, the ureters drain urine from the kidneys, and the urethra drains urine from the bladder.

closed circulation/open circulation

Both are strategies to circulate fluid within the body. However, closed circulation is found in higher animals and blood is contained in vessels, whereas open circulation is found in insects and does not involve vessels.

Matching

a. (reabsorption)
b. (excretion)
c. (secretion)
d. (concentration)
e. (absorption)

Short Answer

1. Capillary beds allow the exchange of materials from the circulatory system to the cells. These materials include wastes from cellular activities and nutrients like oxygen.
2. Oxygen moves from air to blood by diffusion across the alveolar surfaces of the lungs.
3. The liver extracts nutrients and toxins from the blood leaving the intestines.
4. Secretion is the movement of substances from the capillaries to the kidney tubule. Examples of secreted substances include $H+$ and $K+$.

Multiple Choice

1e, 2a, 3b, 4d, 5c, 6c, 7d, 8c, 9d, 10a, 11d, 12a, 13c, 14e, 15b

BUT, WHAT'S IT ALL ABOUT?

1. **Type of Question**

   Apply your knowledge, what is the function of the circulatory system (hint, title of this chapter)?
2. **Evidence Needed**

   Role of circulation is to carry O_2 to tissues and waste products away.
   a. Fatigue—lack of O_2 to the muscles
   b. Lethargy—lack of O_2 to the brain
   c. Infections–lack of O_2 to immune cells, less cellular function because energy not made (remember respiration, Chapter 7?).
   d. Cold feet—again energy, we lose energy as heat during metabolism, which makes our blood warm. If the blood doesn't get to the toes, they stay cold.

Chapter 29

General Processes in Development

Think about It

1.

Name of Stage	Description	Function
Cleavage	Rapid division of the zygote	Formation of the blastula into many cells
Gastrulation	Cell movements and migrations	Creates the three germ layers
Organogenesis	CNS forms by rolling of a tube of ectoderm	Formation of specialized organs

2. C bunny fur and your teeth
   N fish guts and calf liver
   M body-builder muscles and your brother's skull
3. a. cleavage
   b. no
   c. Yes, each piece (cell) gets smaller.

FACTORS UNDERLYING DEVELOPMENT

1. Please refer to the text.
2. Morphogens are diffusible substances whose concentration affects developmental pathways at localized regions. They work by turning on and off specific genes.
3. Induction. Some cells direct the developmental pathway of other cells.
4. a. totipotential → determination → commitment
   b. It is possible to predict its developmental fate.
5. Dolly was cloned from a mature (committed) cell.

SHAPING THE BODY

1. because it is necessary to establish the three germ layers, which involves cells moving relative to one another and moving against or along one another
2. a. gastrulation
   b. adherence/adhesion
3. due to the continuing interactions between genes, proteins, and the environment
4. "baby teeth" falling out and being replaced with mature teeth; puberty; menopause

Vocabulary Review

Cleavage	Gastrulation	Organogenesis
zygote	endoderm	notochord
blastula	invagination	somites
blastocoel	primitive gut	neural tube
animal pole	mesoderm	neural crest
	ectoderm	

PRACTICE TEST

Compare and Contrast

mesoderm/ectoderm

Both are germ layers of the gastrula. However, the mesoderm is the central layer while the ectoderm is the outer layer; and each layer develops into different final tissues in the adult organism.

notochord/neural tube

Both are specific structures formed during early organogenesis. However, the notochord is a flexible rod that induces the formation of the neural tube from ectoderm, whereas the neural tube ultimately forms the CNS.

blastula/gastrula

Both are intermediate structures formed during early development. However, the blastula is a hollow ball of cells, while the gastrula is a three-layered embryo that develops from the blastula.

determined cells/committed cells

Both refer to developmental stages of a cell with respect to its ultimate fate. However, determined cells have a recognizable fate, while committed cells have a fate that cannot be reversed.

Matching

a. (morphogen)
b. (induction)
c. (fertilization)
d. (organogenesis)
e. (gastrulation)

Short Answer

1. Induction is the capacity of some embryonic cells to control the development of others.

2. Stem cells are blastocyst-stage cells that are capable of giving rise to almost any cell in the body.
3. A zygote becomes an embryo after it begins cell division.
4. Morphogens are substances that can diffuse through the embryo, influencing development by their relative concentration.

Multiple Choice

1e, 2b, 3c, 4c, 5e, 6d, 7d, 8c, 9d, 10d, 11c, 12a

BUT, WHAT'S IT ALL ABOUT?

1. **Type of Question**
   Apply your knowledge to solve a question
2. **Evidence Needed**
   "darned if I know" is not an acceptable answer.
   Phylogeny
   Simple organisms are ancestral to complex ones
   Complexity involves gaining specialized structures for new functions
   Controlled by genes (mutations are fuel for natural selection)
   Ontogeny
   Early stages produce general structures
   Later stages produce more complex structures with specific functions from the simpler structures
   Directed by gene products.

Chapter 30

Overview of Human Reproduction

Think about It

1. development of gametes and delivery of gametes to the site of fertilization
2. Testosterone is produced in the testes, and estrogen and progesterone are produced by the ovaries.

THE FEMALE REPRODUCTIVE SYSTEM

1. ovary; uterine tube
2. in the uterine tube
3. It is the lining of the uterus, and supports the development of an embryo if conception and implantation occur.
4. the shedding of the endometrium that is not needed if an embryo does not implant
5. to provide hormones to support a pregnancy
6. fraternal twins — 2 zygotes
   identical twins — 1 zygote, which subsequently splits
7. Fertilization of two eggs that were ovulated at the same time. Mary is more likely to have twins, because she has the capacity to ovulate (and thus the chance to ovulate more than one oocyte per cycle).

8. Please refer to the text.
   1. uterine tube
   2. zygote
   3. oocyte
   4. accessory cells
   5. ovary
   6. uterus
   7. implantation

THE MALE REPRODUCTIVE SYSTEM

1. because they don't have a limited number of gametes, but generate fresh gametes continuously
2. approximately 250 million (250,000,000)
3.

Structure	Function
Testes	Site of sperm development
Epididymis	Sperm maturation and storage
Seminal vesicle	Supply nutrients and lubrication to the sperm
Prostate gland	Supply nutrients and lubrication to the sperm
Urethra	Delivery of sperm
Bulbourethral gland	Supply nutrients and lubrication to the sperm

4. The acrosome contains and releases digestive enzymes so that the sperm can get through the layers surrounding the egg and penetrate the egg.
5. The membrane potential of the egg changes after fertilization, so that a second sperm cannot fertilize the egg. This is important because fertilization of the egg by more than one sperm would result in polyploidy and probable miscarriage.

HUMAN DEVELOPMENT AND BIRTH

1. up to the end of the first trimester, the time during which major developmental events (including organogenesis) occur
2. inner cell mass → embryo
   trophoblast → placenta
3. Surfactant is a lung lubricant. It helps keep the lungs from collapsing. Surfactant is not produced until the 28th week, so premature babies born earlier than that time are at risk for lung problems.
4. 5 days → blastula formation
   7–10 days → implantation
   16 days → gastrulation
   4th week → organogenesis
5. uterine contractions, cervical dilation, and the birth of the baby
6. The afterbirth is the placenta, which is expelled after the baby is delivered.

Vocabulary Review

Female Reproductive System	Male Reproductive System	Human Development
oocyte	spermatocytes	embryo
ovary	seminiferous tubules	zygote
	semen	conception
		placenta
		blastocyst

PRACTICE TEST

Compare and Contrast

sperm/semen

Both are components of the male ejaculate. However, sperm are the male gametes, and semen is the fluid containing sperm and other compounds (nutrients and lubrication).

uterine tube/urethra

Both are structures associated with the human genitourinary tract. However, uterine tubes are found only in females. Both males and females have a urethra, to discharge urine (and sperm in males).

epididymis/corpus luteum

Both are structures in the human reproductive anatomy. However, the epididymis is a site of sperm storage within the testes, while the corpus luteum is the structure remaining after the ovulation of an oocyte from the follicle.

oocyte/zygote

Both are cells involved in development. However, an oocyte is the female gamete, while a zygote is the fertilized (and therefore diploid) egg.

Matching

a. (oogenesis)
b. (atresia)
c. (ovulation)
d. (conception)
e. (bulbourethral gland)

Short Answer

1. Atresia is the natural degeneration of follicles.
2. Internal body temperatures are too high for sperm viability.
3. Identical twins are the result of one egg and one sperm, while fraternal twins require two eggs and two sperm.
4. Prior to 28 weeks of gestation, surfactant is not produced, so the lung tissue sticks to itself when preemies try to breathe.

Multiple Choice

1c, 2c, 3e, 4b, 5d, 6c, 7d, 8d, 9b, 10d, 11c, 12b

BUT, WHAT'S IT ALL ABOUT?

1. **Type of Question**
   Apply your knowledge to answer a question.
2. **Evidence Needed**
   The only evidence we have is the observation that the embryo is not attacked by the maternal immune system (in most cases). Scientists are only beginning to approach this question experimentally as we learn about the immune system in greater detail. We can't answer this question directly, but we can propose some testable hypotheses to explain this observation, just as a practicing scientist might.
   Hypothesis 1: The embryo produces something that blocks the maternal immune system from recognizing non-self antigen, so we could look for novel chemicals in fetal blood.
   Hypothesis 2: The embryo produces something that inactivates the maternal immune system, so we would look for an inhibitor of immune function. If this was happening we might expect that pregnant women would get sick more often than non-pregnant ones.
   Hypothesis 3: The maternal immune system shuts down part of its immune system, like B-cell surveillance, to protect the embryo, so we might find that pregnant women might have fewer circulating B cells than non-pregnant women.
   Hypothesis 4: The embryonic blood never directly contacts maternal blood. Immune cells are too large to diffuse through the membranes so the maternal immune system never sees antigens from the embryo. In this case we should look for immune cells in the "wrong" place (fetal cells in maternal circulation).

Chapter 31

Population Dynamics

Think about It

1. From the outer (largest) to inner (smallest) circle: biosphere, ecosystem, community, population
2. a. the students in your biology class P
   b. the birds, bees, and flowers in your garden C
   c. the soil, fertilizer, and crops in a farmer's field E
   d. the pumpkinseed sunfish in a pond P
   e. the pumpkinseed sunfish, plankton, and plants in a pond C
3. (i) count (number of individuals)
   (ii) distribution of individuals over a geographical area
   (iii) changes in the size of a population over time

4. a. Estimate by counting number in a defined sample volume, and extrapolate to the total volume of the lake.
   b. Estimate by counting number in a defined area, and extrapolate to the total area of the desert.
   c. direct count
   d. direct count
   e. Estimate by counting bear spoor in a defined area, or by monitoring evidence of bear activity (scratches on tree trunks).
5. Please refer to the text.

exponential	fruit flies
arithmetic	number of biology textbooks leaving the printing press

6. a. Please refer to the text for (exponential) and (logistic) curves. Limitation of population growth by environmental factors (environmental resistance) in the case of logistic growth. Examples would include depletion of resources and increase in the number of predators.
   b. logistic growth; carrying capacity
7. Birth rate $= (350/3000) \times 100 = 11.67\%$
   Death rate $= (285/3000) \times 100 = 9.5\%$
   $r =$ birth rate–death rate $= 11.67 - 9.5 = 2.17\%$ per year.
8. No. K (carrying capacity) depends on the specific environmental conditions, which can influence things like resource availability.
9. because it serves as an appropriate yet translatable model
10. Exponential growth is more likely to occur because each individual can produce more offspring, so with each generation the growth of the population is dependent on the number of individuals present in the preceding generation.
11. Overall, the population increased by 7 individuals, which is 7% of the population size.
12. 24 chicks (12% of 200 individuals)
13. rapid growth with minimal investment
14. a. disease (yes)
    b. decrease in temperature (no)
    c. increase in food supply (yes)
15. Please refer to the text.
    r-selected = early loss (insects)
    K-selected = late loss (humans)
16. r- dandelions that move into an untreated lawn and rapidly take over
    K- beech trees that don't reproduce until age 20
    K- elephants, which produce only one offspring at a time
    r- cockroaches (with egg cases full of 100 eggs each)
17. a. Please refer to the text, (Kenya) for a pre-reproductive age population, and (USA) for a post-reproductive age population.
    b. The population with most of its members at or prior to reproductive age is more likely to grow rapidly, because there will be more individuals reproducing.
18. immigration and emigration
19. Both factors are likely to be important in the survival of our species, because both have an impact on the resource utilization and accumulation of pollutants.
20. food, resources for shelter, resources for energy

SPECIES INTERACTIONS WITHIN THE ECOLOGICAL COMMUNITY

Think about It

1. the most abundant species in a community (those with the greatest population numbers)
2. diversity of species; distribution of individuals of species; genetic diversity within a species
3. Genetic diversity will likely suffer, which means that the restored population will have fewer alleles at each locus and less variability to deal with changing environmental conditions.
4. It will depend on the community you have drawn, but the keystone species will be the one that influences the population size of all the others.
5. to understand the interactions of populations and how species share resources
6. a. In experiment 1, A and B are experiencing competitive exclusion (B is outcompeting A for resources). In experiment 2, both C and D are coexisting.
   b. A and B
7. parasitism, because the host survives
8. Mullerian
9. Batesian
10.

predation	parasitism	Batesian mimicry	Mullerian mimicry	commensalism
Both species hurt				**Both species helped**

11. They can allocate the resources between them.

SUCCESSION IN COMMUNITIES

Think about It

1. primary succession
2. P weeds growing through the cracked asphalt of an abandoned gas station
   C a mature, stable forest community
   S a farm field allowed to return to the wild
   P a plant community growing on the cooled lava of a Hawaiian volcano
3. stable, mature communities at the end of the successional process

Vocabulary Review

Community	Population	Ecology
groups of species	biosphere	biodiversity
keystone species	*r*	study of the home
	J-curve	ecosystem
	K	
	groups of individuals	
	S-curve	
	intrinsic rate of growth	
	carrying capacity	

PRACTICE TEST

Compare and Contrast

competition/predation

Both are ways in which populations can interact. However, competition is when individuals attempt to use the same resources, while predation is the consumption of one species by another.

primary/secondary succession

Both refer to ways in which communities change over time. However, primary succession is the establishment of an initial community in an area with no previous living organisms, while secondary succession is the change of the community as it becomes established.

community/population

Both refer to ways in which biological organisms interact. However, populations are collections of interacting individuals, and communities are collections of interacting populations.

mutualism/commensalisms

Both refer to specific interactions between individuals. However, mutualism is a situation in which both individuals interact, while in commensalisms one individual benefits and the other is unaffected.

Matching

a. (competitive exclusion)
b. (mimicry)
c. (ecology)
d. (logistic growth)
e. (coevolution)

Short Answer

1. curves generated by life-table data that depict the probability of mortality for an individual over time
2. commensalism
3. The common processes are facilitation of growth by earlier species for later ones and competition among later species.
4. Parasitism is a variety of predation in which the predator feeds on prey, but is unlikely to kill it.

Multiple Choice

1d, 2c, 3d, 4c, 5e, 6c, 7e, 8c, 9d, 10c, 11c, 12c

BUT, WHAT'S IT ALL ABOUT

1. **Type of Question**
   Defend a position. Given what you know about viruses and the human body from chapter 27, can this relationship be described as parasitic?
2. **Evidence Needed**
   Make a table!

Yes, it is parasitic	No, it is not parasitic
Viruses benefit, hosts do not.	Hosts do benefit by developing antibodies & T-cells against that virus, providing long term protection.
Viruses compromise host functioning by sucking up resources.	Viruses do not infect hosts for nourishment, they aren't alive.
Viruses can kill their host.	By weakening or killing the host, the virus would fail to guarantee its own survival.
Viruses can evolve to avoid the host immune system.	

Chapter 32

Ecological Systems

Think about It

1. a. A
   b. A
   c. B
   d. B
2. Ecosystems include abiotic components of the physical world; communities are interactions of living populations only.
3. carbon, oxygen, nitrogen, phosphorus
4. burning of fossil fuels (coal and oil)
5. Plants generate/<u>consume</u> CO_2 by the process of <u>photosynthesis</u>, and animals <u>generate</u>/consume CO_2 through the process of <u>cellular respiration</u>.
6. through the process of decomposition by soil bacteria and fungi
7. proteins and nucleic acids
8. a. Starting material—atmospheric nitrogen; End product—ammonia
   b. It is the only natural source of fixed (and therefore useable by organisms) nitrogen.
9. a. evaporation from the oceans and lakes; transpiration from plants
   b. The Sun is critical for both evaporation and transpiration. It is the Sun's energy that essentially drives these processes.
10. fertilizer and nitrogen-fixing bacteria

ENERGY FLOW THROUGH ECOSYSTEMS

Think about It

1. a. ordered
   b. ordered
   c. disordered
   d. ordered
2. a. 4th-coyote, 3rd-house cat, 2nd-blue jay, 1st-sunflower
   b. at least 10 (assuming the predator is no larger than the prey animals)
3. the ability to degrade man-made materials, such as plastics

4. a. Approximately 10% of the energy of each trophic level is converted to biomass in the next trophic level.
   b. This value is low due to much of the biomass being in a given level being inedible by organisms in the next level, and due to the energy required to maintain the body before any additional biomass can be accumulated.
5. a. a gas consisting of three oxygen molecules bonded together
   b. stratosphere
   c. protection from mutagenic UV irradiation
   d. CFCs destroy the protective ozone layer, so that more mutagenic (and therefore cancer-causing) UV irradiation reaches our skin cells.
6. The accumulation of CO_2 helps trap energy in the form of heat within the Earth's atmosphere.
7. a. Please refer to the text.
   b. Please refer to the text.
   c. because the Sun's rays hit the Arctic at a more oblique (less direct) angle than they hit Florida
8. The ocean side will be relatively wet. This is because the wet air is forced to move up to cross the mountain. The increase in altitude causes the air to cool, and to lose its ability to hold moisture. Thus rain falls before the air crosses the mountain. Once it crosses, it has very little moisture.
9. The source is sunlight, which hits the equator with the most intensity, and then sets up the global convection of air currents.

EARTH'S BIOMES

1. (1) tundra a
   (2) taiga e
   (3) temperate deciduous forest b
   (4) grassland f
   (5) chaparral g
   (6) deserts c
   (7) tropical rain forest d
2. Please refer to the text.
   1. intertidal zone
   2. coastal zone
   3. open sea
   4. photic zone
   5. pelagic zone
   6. benthic zone
3. phytoplankton (producers) and zooplankton (consumers)
4. Bacteria and algae will begin to divide due to the presence of additional nutrients; as they grow, they consume oxygen in the process of cellular respiration. The resulting depletion of oxygen from the water causes the fish to asphyxiate.

Vocabulary Review

Biomes	Nutrient Cycling	Climate
tundra	fossil fuels	rain shadow
taiga	decomposition	fossil fuels
temperate forest	nitrogen	stratosphere
tropical rain forest	carbon	greenhouse effect
chaparral	nitrogen-fixing bacteria	

PRACTICE TEST

Compare and Contrast

herbivore/detritivore

Both are consumers in a food chain or food web. However, herbivores eat producers (plants), while detritivores consume dead organic material, serving as the recyclers.

gross primary production/net primary production

Both refer to conversion of the Sun's energy to plant biomass by photosynthesis. However, gross primary production is the total amount of energy assimilated, while net primary production is the total amount of energy assimilated from which the amount of energy required to maintain the plant is subtracted.

estuary/wetland

Both are bodies of water. However, estuaries are the areas where streams or rivers flow into the ocean, while wetlands are wet (submerged) for a portion of the year.

ecosystem/biome

Both refer to ecological levels of organization. However, ecosystems include the organisms and their biotic and abiotic environments, while biome is a community defined by the vegetation.

Matching

a. (carbon dioxide acts as an insulator in the Earth's atmosphere)
b. (photosynthesis)
c. (conversion of carbon dioxide to carbohydrate)
d. (change in the profile of a lake due to increased nutrient amounts)
e. (formation of new desert areas)

Short Answer

1. Most of the nutrients in the rain forest biome are held within the living tissue (biomass) instead of the soil. The opposite is true in a deciduous forest.
2. More carbon is available to plants in the atmosphere through the burning of wood, coal, and fossil fuels.

3. The two types of factors that help to mold ecosystems are conditions and resources.
4. The concentration of heat-energy in the atmosphere due to the increase in greenhouse gases, which prevent the energy from radiating to outer space. This in turn causes an increase in temperature.

Multiple Choice

1d, 2b, 3b, 4d, 5e, 6c, 7a, 8a, 9d, 10d, 11c, 12b, 13a

BUT, WHAT'S IT ALL ABOUT

1. **Type of Question**
   Apply your knowledge to identify essential components of a self-contained earth environment
2. **Evidence Needed**
   Identify and describe the challenges. You might want to do some web searching and learn about Biosphere 2, a 3.1 acre, glass enclosed facility that was the site of two experiments with humans living in a self-contained environment.

   Regulation of the carbon cycle—Need to provide controlled environment (temperature, humidity, soil nutrients) for the plants so that the amount of CO_2 produced by the non-plant organisms is sequestered appropriately. Too much CO_2 will make people lethargic, too little would inhibit photosynthesis, compromising the food supply. Also need to worry about the plants making enough O_2.

   Regulation of the hydrologic cycle—Everything inside the community needs water, but the water vapor doesn't have a near-infinite space to expand into. Excess water collecting on surfaces would favor mold and bacterial growth—not a favorable human environment. The consequences of a lack of water are obvious.

   Nitrogen cycle—Dealing with waste and decay is one of the hardest issues to resolve. Need to include nitrogen-fixing plants in the mix, but also need to worry about excess nitrates in the soil.

Chapter 33

Behavioral Biology

Think about It

1. We are usually aware of our proximate drives. Ultimate causation (evolutionary success) tends to be less straightforward when examining human behavior.
2. The biologist changes some one thing in the environment and observes the effect of the change upon the animals' behavior. With repeated experiments, they will be able to determine the proximate cause of the behavior.
3. Animals do not choose their actions, because choosing implies being able to predict the short-term and long-term effects of selecting between equivalent behavior patterns. Most animal behavior is the result of natural selection; behaviors that have improved the survival or reproductive success of the species in the past have been "passed down" to later generations. Animals can learn new behaviors if those behaviors provide an immediate benefit to the animal.

INFLUENCES ON BEHAVIOR

1. Some situations require nothing more than a reflex response to a stimulus. Others require that the response be sculpted by various factors. This last would require learning and perhaps insight.
2. Environment provides sensory input to either trigger or modify behavior.
3. Action patterns are triggered by an internal or external stimulus and always proceed to completion without modification. Learned behavior may be triggered by a stimulus but can be modified based on experience.
4. a. Environmental cues can signal the start of a circadian cycle
   b. Not necessarily, because human circadian rhythms are partially entrained by light/dark cues, we would expect a blind person's circadian rhythms to be different from those of a sighted person. But other, internal cues also affect human rhythms, so the variation in cycle length might be very small.
5. Angela had been habituated to warmer temperatures and reacted to the colder climate by changing her clothing choice behavior until she could habituate to the new, lower, temperature.
6. Classical conditioning—Fluffy associates the sound of the motor with being fed.
7. Operant conditioning—Fluffy has learned that interrupting your dinner means she spends time in the dreaded box, so to avoid the box she stops jumping onto the table.

SOCIAL BEHAVIOR

1. Probably all would be considered social species, although cats less so than the others.
2. Lekking is a dominance hierarchy, although protecting territory is part of the process. Some males in the group dominate the others by securing and keeping the best territories in the lek and the largest number of mates.
3. a. My daughter shares 50% of her genes with me, her father, and her siblings. She shares about 25% of her genes with my older sister's children, and about 12.5% of her genes with both of my sisters, my older sister's husband, and my younger sister's children. The most likely donors occur within our immediate family; but my older sister's children, who are related to my daughter through both their father and their mother, are the next most genetically similar individuals.
   b. reciprocal altruism

Vocabulary Review

Learning and Internal Influences	Social Behavior
habituation	eusocial
insight	dominance hierarchies
communication	kin selection
imprinting	altruism
action patterns	communication
migration	proximate causes
taxis	ultimate causes
trial-and-error learning	inclusive fitness

Learning and Internal Influences	Social Behavior
orientation imitation reflexes ultimate causes classical conditioning operant conditioning circadian rhythms proximate causes	

Matching

a. (seasonal movement based on an internal clock)
b. (learning by copying another's behavior)
c. (reduction in response based on repeated exposure)
d. (learning that takes place during a sensitive period)
e. (behavior performed at a direct cost to the actor)

Compare and Contrast

proximate/ultimate causation
Both cause a given behavior to occur. However, proximate causation involves stimuli occurring right now, while ultimate causation would be the effect on an individual's fitness in evolutionary terms.

habituation/insight
Both are levels of learning. However, habituation would be one of the simplest levels of learning, while insight is one of the most complex.

taxis/migration
Both involve movement. However, taxis is simple movement toward or away from a nearby stimulus. Migration is quite complex and may involve movement over thousands of miles.

altruism/kin selection
Both are observed in social animals. However, kin selection involves the selection of genes that increases the inclusive fitness of an individual.

Short Answer

1. Behavioral biology is the study of behavior using natural selection as a conceptual framework. As Dobzhansky said, "Nothing in biology makes sense save in the light of evolution."
2. Birdsong can occur naturally without external cues. However, species-specific song and local dialects are the result of learning.
3. Costs—increased visibility to predators, decreased availability of resources (food, etc.), increased likelihood of communicable disease, and so on

Benefits—cooperative behavior (hunting, offspring care), altruistic interaction, defense, and so on

4. Reciprocal altruism involves a trading of altruistic acts. In other words, altruism that is performed with the expectation of reward.

Multiple Choice

1e, 2b, 3d, 4b, 5b, 6e, 7a, 8d, 9c, 10d, 11a, 12c

BUT, WHAT'S IT ALL ABOUT?

1. **Type of Question**—Tricky
2. **Evidence Needed**

   Given the frequency of abandonment and child abuse in our modern world, you could argue that my premise, that humans demonstrate altruism, is false. Yet, if you examine the broader human experience, as described in stories, in rituals, in societal values, you could find the evidence to support the contention that humans are altruistic animals, primarily to their kin but also to strangers. Why? Maybe biology can explain life well and behavior reasonably well, but it needs the help of literature, philosophy, ethics, and the arts to explain love. The natural sciences are only one piece of a liberal arts education—remember to apply the arts and humanities too when solving the truly important questions of this world.